THE PREPPER'S WATER SURVIVAL GUIDE

Essential Skills for Finding, Treating, and Storing Water in Crisis

Jordan Riverdale

Copyright © 2024, Jordan Riverdale. All rights reserved.

No part of this publication may be reproduced, distributed, or transmitted in any form or by any means, including photocopying, recording, or other electronic or mechanical methods, without the prior written permission of the publisher, except in the case of brief quotations embodied in critical reviews and certain other noncommercial uses permitted by copyright law.

The copyright for this book, "THE PREPPER'S WATER SURVIVAL GUIDE," includes but is not limited to the following:

- The text, narratives, descriptions, instructions, and any other written content.

- Any derivative works based on the original, including translations, adaptations, summaries, and multimedia formats.

- Images, charts, diagrams, illustrations, and visual content specifically created for this work.

- The cover design, graphic elements, layout, and typography.

- Protection of digital rights concerning distribution and reproduction in digital form.

- The book's title and any distinctive elements used as trademarks for identifying the source of the book.

- Rights of publication and distribution, specifying agreements with publishers, distributors, and sales platforms.

- Attribution and integrity clauses to ensure proper author attribution and to protect the work's integrity against unauthorized alterations.

This copyright notice does not grant any right to use the material contained in "THE PREPPER'S WATER SURVIVAL GUIDE" beyond what is allowed by copyright law without the prior written consent of Jordan Riverdale or the legal permission obtained for the use of any material that is not owned by the author.

Table of Content

Table of Content..5

Introduction..9

Understanding the Importance of Water................11

    1.1 The Vital Role of Water in Survival...............12

    1.2 Water Sources....................................14

    1.3 Water Contamination.........................20

    1.4 Water Storage..................................25

    Conclusion..29

Mastering Water Treatment Techniques...................31

    2.1 Boiling Water....................................32

    2.2 Chemical Treatment.........................35

    2.3 Filtration Systems.............................41

    2.4 Solar Disinfection..............................46

Rainwater Harvesting and Collection........................52

    3.1 The Basics of Rainwater Harvesting...............53

    3.2 DIY Rainwater Collection Systems...............56

3.3 Rainwater Purification..58

3.4 Creative Uses of Rainwater............................62

3.5 Legal Considerations...67

Water Conservation and Efficiency...........................72

4.1 The Importance of Water Conservation in Crisis...73

4.2 Reducing Water Waste....................................76

4.3 Greywater Recycling..80

4.4 Xeriscaping...84

Emergency Water Sources...90

5.1 Natural Water Sources....................................91

5.2 Urban Water Sources......................................94

5.3 Alternative Water Sources.............................98

5.4 Water from Unconventional Sources...........102

Water Storage and Containers....................................108

6.1 Choosing the Right Water Storage Containers 109

6.2 Water Storage Guidelines.............................113

6.3 Water Rotation...117

6.4 Emergency Water Storage Solutions............122

Water for Hygiene and Sanitation............................128

   7.1 Maintaining Personal Hygiene in Water-Scarce Situations...129

   7.2 Sanitation Practices...131

   7.3 Waterless Hygiene Alternatives......................136

   7.4 Emergency Toilet Options..............................141

Water in Extreme Environments...............................146

   8.1 Surviving in Arid and Desert Environments 147

   8.2 Water Challenges in Cold Climates...............149

   8.3 Water in Coastal and Marine Environments 154

   8.4 Water in Urban Environments......................158

Water for Food Production..163

   9.1 Irrigation Techniques......................................164

   9.2 Aquaponics...169

   9.3 Watering Livestock..172

   9.4 Preserving Food with Water..........................176

Preparing for Long-Term Water Scarcity...............181

   10.1 Drought Preparedness..................................182

   10.2 Water Conservation in the Long Run........184

   10.3 Community Water Solutions.......................187

   10.4 Planning for Water Security........................193

Conclusion..201

Introduction

In a world teetering on the brink of instability, water remains the silent sentinel of survival. For the prepper, it is not merely a resource; it is the lifeblood of preparedness. In the parched landscape of a crisis, where each drop counts, the knowledge to harness, purify, and store water can mean the difference between thriving and merely surviving.

This book, "The Prepper's Water Survival Bible," is your guide through the treacherous terrain of water scarcity and uncertainty. It's an armamentarium of wisdom, tactics, and techniques that will prepare you for the worst—because when society's structures falter, the flow of water does not wait for resolutions.

As you turn these pages, you will learn to look at water with a new reverence. You will understand why every prepper must prioritize water above all else and how to assess your water vulnerability accurately. With this bible in hand, you will never underestimate the power of being water-wise.

You will embark on a journey through the art of hydration, from uncovering hidden sources to creating fortresses of rainwater. You will become adept at deciphering the signs of water in the wild, mastering the methods of making any water drinkable, and learning the critical techniques of guarding your precious reserves.

This is not just about being prepared—it is about being strategically prepared. It is about ensuring that, even when the world around you is parched, you have the knowledge and the power to conjure water, to purify it, and to protect it.

Let's begin this vital journey with a sobering exploration of what life could look like without this crucial resource. It's a glimpse into the stark realities that await the unprepared, and a beacon of knowledge for those who choose to take the prepper's oath: to remain water aware and prepared, no matter what.

To the prepper, water is not just a resource; it's a responsibility. Welcome to "The Prepper's Water Survival Bible." Your path to becoming a water-wise warrior starts now.

Understanding the Importance
of Water

1.1 The Vital Role of Water in Survival

Water is the essence of life. It is the elixir that sustains us, nourishes us, and keeps us alive. In our everyday lives, we often take water for granted, assuming it will always be readily available at the turn of a faucet. But what happens when that faucet runs dry? What happens when a crisis strikes and access to clean water becomes uncertain?

As a prepper, you understand the importance of being prepared for any emergency. And when it comes to survival, water should be at the top of your priority list. Without water, our bodies cannot function properly. It regulates our temperature, aids in digestion, carries nutrients to our cells, and flushes out waste. In a crisis situation, the availability of clean water can mean the difference between life and death.

1.1.1 The Consequences of Dehydration

Dehydration is a serious threat to our well-being, and its effects can be devastating. Even mild dehydration can lead to fatigue, dizziness, and confusion. In more severe cases, it can cause organ failure and even death. When faced with a water shortage, it is crucial to understand the signs of dehydration and take immediate action to prevent it.

1.1.2 Water as a Basic Human Right

Access to clean water is not only essential for our individual survival but also a basic human right. Unfortunately, in times of crisis, water sources can become contaminated or inaccessible. Natural disasters, such as hurricanes or earthquakes, can disrupt water infrastructure, leaving communities without a reliable supply. By being prepared and knowledgeable about water survival techniques, you can ensure that you and your loved ones have access to this fundamental resource.

1.1.3 The Importance of Water in Emergency Situations

In emergency situations, water serves multiple purposes beyond hydration. It is crucial for cooking, cleaning, and maintaining personal hygiene. Without clean water, the risk of waterborne diseases increases significantly. By understanding the vital role of water in survival, you can better prepare for emergencies and ensure the well-being of yourself and your family.

1.1.4 The Psychological Impact of Water Scarcity

Water scarcity not only affects our physical health but also takes a toll on our mental well-being. The stress and anxiety of not having access to clean water can be overwhelming. By being prepared and equipped with the knowledge and skills to handle water-related emergencies, you can alleviate some of this psychological burden and provide a sense of security for yourself and your loved ones.

1.1.5 The Prepper's Responsibility

As a prepper, you have taken on the responsibility of safeguarding yourself and your family in times of crisis. By understanding the vital role of water in survival, you are one step closer to being fully prepared. In the following chapters, we will delve deeper into the various aspects of water survival, from finding and treating water sources to storing water for the long term.

So, let's dive in and explore the essential skills for finding, treating, and storing water in crisis. Remember, water is life, and by being prepared, you are ensuring the survival and well-being of yourself and your loved ones.

1.2 Water Sources

Water is the essence of life. It sustains us, nourishes us, and keeps us alive. In a crisis situation, access to clean and safe water becomes even more critical. As a prepper, it is essential to understand where to find water when traditional sources are compromised. In this chapter, we will explore various water sources that can be relied upon during emergencies and how to make the most of them.

1.2.1 Natural Water Sources: Rivers, Lakes, and Streams

When disaster strikes, the significance of natural water sources such as rivers, lakes, and streams becomes paramount. These sources, abundant in many areas, can serve as a lifeline, offering a steady supply of essential water, provided they are properly accessed and treated. To fully harness these natural reserves, it's vital to master several techniques for locating and accessing them, ensuring a sustainable water supply during emergencies.

First and foremost, understanding the landscape and environmental signs is key to identifying potential water sources. Observing the flight patterns of birds early in the morning or late in the afternoon can lead to water, as can tracking animal tracks, which often converge towards rivers, lakes, or streams.
Vegetation is another telltale sign; areas with lush, green vegetation are likely to be near a water source.

Additionally, low-lying areas and valleys are natural collection points for water and should be explored.

1.2.2 Urban Water Sources: Tapping into Municipal Systems

In urban environments, the labyrinth of municipal water systems forms the lifeline of our daily water supply. Yet, during emergencies, these complex systems are vulnerable to disruptions, potentially leaving countless residents facing a dire lack of access to clean, potable water. As a prepper, acquiring the knowledge and skills to safely and efficiently tap into these systems can make all the difference.

Understanding the infrastructure of municipal water systems is the first step. This includes knowing the locations of primary water lines, shut-off valves, and fire hydrants, which can become crucial access points in times of need. However, accessing these points requires specific tools and precautions. For instance, a water key or a four-way silcock key is essential for opening commercial water access points found on the exterior of many buildings or in public spaces.

When primary systems fail, knowing how to identify alternative water sources within an urban setting becomes invaluable. Beyond the obvious lakes and rivers, other sources include rainwater collection, condensation from air conditioning units, or water heaters and toilet tanks (not the bowl) within buildings, which can store potable water. However, it's crucial to apply the same purification standards to these sources as well.

1.2.3 Alternative Water Sources: Ponds, Wells, and Cisterns

In scenarios where traditional water sources become scarce or contaminated, turning to alternative sources like ponds, wells, and cisterns can offer a lifeline. These sources, while requiring more effort to access and treat, present a crucial reservoir of water during emergencies. Understanding the nuances of each source is essential for ensuring the safety and sustainability of your water supply.

Ponds offer a natural collection of surface water that can be tapped into when other sources are unavailable. However, the static nature of pond water means it's more susceptible to contamination from environmental runoffs, animal waste, and microbial growth. To safely use pond water, it's crucial to implement robust filtration and purification methods, such as sediment filtration followed by boiling or chemical treatment.

Wells, tapping into underground aquifers, can provide a cleaner and more reliable water source. The depth of the well plays a significant role in water quality, with deeper wells generally offering safer water. However, the process of accessing well water typically requires mechanical means, such as a hand pump or an electric pump. It's vital to regularly test well water for contaminants like heavy metals, bacteria, and chemicals, particularly in areas near agricultural or industrial activities.

Cisterns are storage systems designed to collect and store rainwater or water from other sources. While they offer a method to harness and control your water supply, the quality of water in cisterns heavily depends on the cleanliness of the collection system and the storage conditions. Regular maintenance to prevent algae growth and contamination, along with thorough filtration and purification before use, is essential.

1.2.4 Water from Unconventional Sources: Dew, Plants, and More

In dire situations where traditional water sources are scarce or inaccessible, it becomes necessary to think outside the box. Nature provides us with unconventional sources of water that, with the right knowledge and techniques, can be utilized for survival.

Dew Collection: Dew forms on surfaces in the early morning or late evening, particularly in humid environments. This can be collected using absorbent materials like cloths or tarps laid out overnight. Once the material is saturated, it can be wrung out into a container. Though labor-intensive and yielding relatively small amounts, this method can accumulate enough water for basic survival needs, especially when other sources are not available.

Water from Plants: Certain plants can be a source of hydration. Techniques such as the solar still method, where a plastic sheet is used to cover a leafy branch or a hole in the ground containing vegetation, can condense and collect water. Additionally, knowledge of specific plants that store water in their tissues, such as cacti or vines, can be crucial. However, it's vital to have a good understanding of local flora to avoid toxic plants.

Understanding where to find water in a crisis is a fundamental skill for any prepper. By exploring natural water sources, urban water systems, alternative sources, and unconventional methods, you will be better equipped to handle water-related emergencies. Remember, water is life, and being prepared to find and treat it can make all the difference in a survival situation.

1.3 Water Contamination

Water contamination is a serious concern in any crisis situation. When disaster strikes, the safety of our water sources can be compromised, leading to the presence of harmful contaminants that can cause illness or even death if consumed. As a prepper, it is crucial to understand the risks associated with water contamination and to know how to identify and treat waterborne hazards effectively.

1.3.1 Understanding Water Contamination

Water contamination occurs when harmful substances enter our water sources, making them unsafe for consumption. These contaminants can come from various sources, including industrial waste, agricultural runoff, sewage, and natural disasters such as floods or earthquakes. Understanding the different types of water contamination is essential for effectively treating and purifying water in a crisis.

Microbial Contamination

Microbial contamination refers to the presence of bacteria, viruses, and parasites in water. These microorganisms can cause diseases such as cholera, dysentery, and giardiasis. They are often found in water sources contaminated with human or animal waste. It is crucial to be able to identify signs of microbial contamination and take appropriate measures to treat the water before consumption.

Chemical Contamination

Chemical contamination occurs when harmful chemicals, such as pesticides, heavy metals, or industrial pollutants, enter the water supply. These chemicals can have severe health effects, including organ damage, cancer, and developmental issues. Identifying signs of chemical contamination and knowing how to remove or neutralize these chemicals is vital for ensuring the safety of your water.

Radiological Contamination

Radiological contamination refers to the presence of radioactive substances in water. This type of contamination can occur during nuclear accidents or incidents involving radioactive materials. Exposure to radioactive substances can lead to radiation sickness, cancer, and other serious health problems.

Understanding the risks associated with radiological contamination and knowing how to protect yourself and treat contaminated water is crucial in a crisis situation.

1.3.2 Identifying Waterborne Hazards

Identifying waterborne hazards is the first step in effectively treating contaminated water. By understanding the signs and symptoms of water contamination, you can take appropriate measures to ensure the safety of your water supply. Here are some key indicators of waterborne hazards:

Unusual Color, Odor, or Taste

Water that appears cloudy, has a strange odor, or tastes unusual may indicate the presence of contaminants. These signs can be an indication of microbial, chemical, or radiological contamination. If you notice any of these characteristics in your water, it is essential to treat it before consumption.

Presence of Sediment or Particles

Visible sediment or particles in water can be a sign of contamination. These particles can come from various sources, including soil, rusted pipes, or industrial waste. Filtering or purifying water with visible sediment is crucial to remove any potential contaminants.

Reports of Contamination in the Area

Staying informed about local news and reports of water contamination in your area is essential. Government agencies and local authorities often issue warnings or advisories when water sources are compromised. Paying attention to these reports can help you take necessary precautions and treat your water accordingly.

1.3.3 Treating Waterborne Hazards

Treating waterborne hazards is crucial to ensure the safety of your water supply. There are several methods available for treating contaminated water, depending on the type and severity of the contamination. Here are some common treatment techniques:

Boiling Water

Boiling water is one of the oldest and most reliable methods of water treatment. By bringing water to a rolling boil for at least one minute (or three minutes at higher altitudes), you can effectively kill most microorganisms, including bacteria and viruses. Boiling does not remove chemical or radiological contaminants, so it is essential to use additional treatment methods if necessary.

Chemical Treatment

Chemical treatment involves using disinfectants or water purification tablets to kill or neutralize microorganisms in water. Common disinfectants include chlorine bleach and iodine. These chemicals can effectively kill bacteria, viruses, and parasites. However, chemical treatment may not be effective against certain types of parasites, such as Cryptosporidium. It is important to follow the instructions provided with the disinfectant or purification tablets for proper use.

Filtration Systems

Filtration systems are designed to remove physical impurities and some microorganisms from water. There are various types of water filters available, including activated carbon filters, ceramic filters, and reverse osmosis filters. Each type of filter has its advantages and limitations, so it is important to choose the right filter for your specific needs. Regular maintenance and replacement of filter cartridges are essential to ensure the effectiveness of the filtration system.

Ultraviolet (UV) Treatment

UV treatment involves using ultraviolet light to kill microorganisms in water. UV water purifiers are portable devices that emit UV light to disinfect water. This method is effective against bacteria, viruses, and parasites. However, it does not remove chemical or radiological contaminants. UV treatment is a convenient and quick method for treating small quantities of water.

1.4 Water Storage

Water storage is a crucial aspect of prepping for any water-related emergency. When a crisis strikes and water becomes scarce or contaminated, having a reliable supply of stored water can mean the difference between survival and desperation. In this chapter, we will explore the essential skills and knowledge needed to safely store water for long-term survival.

1.4.1 The Importance of Water Storage

In a crisis situation, access to clean and safe water may become limited or even non-existent. It is essential to have a sufficient supply of water stored to meet your basic needs, such as drinking, cooking, and personal hygiene. Without proper water storage, you risk dehydration, illness, and overall compromised health.

Water storage is especially critical for preppers because they understand the unpredictable nature of emergencies. Whether it's a natural disaster, a breakdown in the municipal water supply, or a long-term survival scenario, having a well-planned water storage system ensures that you and your loved ones have a reliable source of water when it matters most.

1.4.2 Determining Your Water Storage Needs

Before you start storing water, it's important to determine how much water you and your family will need during an emergency. The general rule of thumb is to store at least one gallon of water per person per day for drinking and sanitation purposes. However, it's always better to err on the side of caution and store more water if possible.

Consider factors such as the number of people in your household, the duration of the emergency, and any specific medical or dietary needs. It's also crucial to account for pets and livestock if you have them. By calculating your water needs in advance, you can ensure that you have an adequate supply to sustain you through the crisis.

1.4.3 Choosing the Right Water Storage Containers

When it comes to storing water, the choice of containers is crucial. You need containers that are safe, durable, and capable of preserving the quality of the water over an extended period. Here are some key considerations when selecting water storage containers:

- **Food-Grade**: Ensure that the containers you choose are made from food-grade materials, such as high-density polyethylene (HDPE) or polypropylene (PP). These materials are safe for storing water and do not leach harmful chemicals into the water.
- **Sealability**: Look for containers with tight-fitting lids or caps to prevent contamination and evaporation. A good seal will also help maintain the water's freshness and prevent the growth of bacteria or algae.
- **Durability**: Opt for containers that are sturdy and resistant to impact, punctures, and UV radiation. This is especially important if you plan to store water outdoors or in areas with fluctuating temperatures.
- **Size and Portability**: Consider the available space in your storage area and the ease of transporting the containers if necessary. Smaller containers may be more manageable for daily use, while larger containers are suitable for long-term storage.

- **Transparency**: Choose containers that are transparent or have a clear strip to allow you to monitor the water's quality and quantity without opening the container.

1.4.4 Proper Water Storage Techniques

Once you have selected the appropriate containers, it's essential to follow proper water storage techniques to ensure the longevity and safety of your stored water. Here are some key guidelines to consider:

- **Clean and Disinfect**: Before filling your containers, thoroughly clean them with mild soap and water. Rinse them well to remove any residue. To disinfect the containers, you can use a solution of one teaspoon of unscented household bleach per gallon of water. Let the solution sit for a few minutes, then rinse thoroughly.
- **Fill and Label**: Fill the containers with clean, potable water from a reliable source. Label each container with the date of filling to keep track of the water's freshness. It's also helpful to label the containers with the amount of water they hold.
- **Store in a Cool, Dark Place**: Find a cool, dark location to store your water containers.

Avoid areas that are exposed to direct sunlight or extreme temperature fluctuations, as these can degrade the quality of the water and the containers.

- **Rotate and Refresh**: Regularly rotate your water storage by using and replacing the oldest containers first. This practice ensures that you always have a fresh supply of water available. Aim to refresh your water storage every six months to a year, depending on the storage conditions and the type of containers used.

By following these water storage techniques, you can maintain a reliable supply of clean and safe water for an extended period. Remember, water is a precious resource, and proper storage is essential for your survival and well-being during a crisis.

Conclusion

In conclusion, the journey through understanding and preparing for water-related emergencies underscores the undeniable importance of water as a fundamental pillar of survival. From recognizing the myriad sources of water — natural, urban, alternative, and unconventional — to addressing the complexities of water contamination and the criticality of proper storage, this comprehensive exploration equips preppers with the knowledge and skills necessary to ensure a sustainable water supply in times of crisis.

The essence of preparedness lies not only in the accumulation of supplies and information but also in fostering a mindset of resilience, adaptability, and community engagement. By embracing the responsibility to safeguard ourselves and our loved ones, we can mitigate the psychological impacts of water scarcity and stand prepared to face the challenges that emergencies present.

As we close this chapter, remember that water is more than a mere resource; it is the lifeline that sustains all aspects of life. The efforts we make today to understand its value, to learn how to source, purify, and store it, are investments in our survival and well-being. Let this knowledge empower you to take action, to be vigilant, and to continuously refine your preparedness strategies. In doing so, you affirm your commitment to not just surviving, but thriving, no matter what challenges the future may hold.

In the next chapter, we will explore various water treatment techniques, including boiling, chemical treatment, filtration systems, and solar disinfection, to further enhance your water preparedness skills.

Mastering Water Treatment Techniques

2.1 Boiling Water

Water is essential for our survival, and in times of crisis, it becomes even more crucial to ensure that the water we consume is safe and free from harmful contaminants. Boiling water is one of the oldest and most reliable methods of water treatment, and it should be a skill that every prepper masters. In this section, we will explore the ins and outs of boiling water, including its benefits, the proper technique, and some practical tips to make the most out of this simple yet effective method.

2.1.1 The Benefits of Boiling Water

Boiling water is a tried and true method of water treatment that has been used for centuries. When you boil water, you raise its temperature to a point where harmful microorganisms, such as bacteria, viruses, and parasites, are killed or inactivated. This makes the water safe to drink and reduces the risk of waterborne illnesses.

One of the greatest advantages of boiling water is its simplicity. All you need is a heat source, a pot, and water. It is a method that can be easily implemented in any situation, whether you are at home, camping in the wilderness, or facing a disaster scenario.

Boiling water requires no special equipment or chemicals, making it accessible to everyone.

2.1.2 The Proper Technique for Boiling Water

To effectively kill or inactivate harmful microorganisms, water must be brought to a rolling boil. This means that large bubbles are rapidly rising to the surface and breaking. It is important to maintain this vigorous boil for at least one minute, or three minutes at higher altitudes where water boils at a lower temperature due to decreased atmospheric pressure.

When boiling water, it is essential to use a clean pot and ensure that the water is free from visible debris. If the water is cloudy or contains sediment, it should be filtered or settled before boiling. Once the water has reached a rolling boil, it is safe to remove it from the heat source and allow it to cool before consumption.

2.1.3 Practical Tips for Boiling Water

While boiling water is a simple process, there are a few practical tips that can help you make the most out of this method:

- **Conserving Fuel**: Boiling water requires a heat source, which may be limited in a crisis situation. To conserve fuel, consider using a lid on your pot to reduce heat loss and boil only the amount of water you need.

- **Storing Boiled Water**: Boiled water can be stored in clean, airtight containers for future use. It is important to label the containers with the date of boiling and to rotate the water regularly to ensure freshness.
- **Enhancing the Taste**: Boiling water can sometimes affect its taste by removing dissolved oxygen. To improve the taste, you can pour the boiled water back and forth between two clean containers to reintroduce oxygen.
- **Using Boiled Water for Cooking**: Boiled water can be used for cooking, making hot beverages, and rehydrating food. It is a versatile resource that can be utilized in various ways during a crisis.

Remember, boiling water is an effective method of water treatment, but it does not remove chemical contaminants or pollutants. If you suspect that your water may be contaminated with chemicals, it is advisable to use additional treatment methods, such as activated carbon filtration or chemical disinfection.

In conclusion, boiling water is a fundamental skill that every prepper should master. It is a reliable and accessible method of water treatment that can be implemented in any situation. By understanding the proper technique and following some practical tips, you can ensure that the water you consume is safe and free from harmful microorganisms. So, grab your pot, fire up your stove, and start boiling your way to water security in times of crisis.

2.2 Chemical Treatment

Water is vital for our survival, and during times of crisis, access to clean and safe water becomes even more important. While boiling water is a reliable method for purifying it, there may be situations where you don't have access to fire or the necessary resources. That's where chemical treatment comes in handy. In this section, we will explore the use of household items to purify water and ensure its safety for consumption.

2.2.1 The Power of Chemicals

Chemical treatment is a simple and effective way to kill harmful microorganisms in water. By using common household items, you can neutralize bacteria, viruses, and parasites that may be present in contaminated water sources. The chemicals work by disrupting the cellular structure of these microorganisms, rendering them harmless.

2.2.2 Common Household Chemicals for Water Treatment

You might be surprised to learn that you already have some of the necessary chemicals for water treatment in your home. Here are a few commonly used household items that can be used to purify water:

- **Bleach**: Regular unscented household bleach can be used to disinfect water. It contains chlorine, which kills most microorganisms. However, it's important to use bleach without any additives or scents, as they can be harmful when ingested. We will discuss the proper bleach-to-water ratio and the steps for using bleach as a water treatment method.
- **Iodine**: Iodine tablets or tincture can be used to disinfect water. Iodine is effective against a wide range of microorganisms, including bacteria and viruses. We will explore the correct dosage and the necessary contact time for iodine to effectively treat water.
- **Calcium Hypochlorite**: Calcium hypochlorite is a powdered form of chlorine

that can be used to disinfect water. It is commonly found in pool shock treatments. We will discuss the proper dilution ratio and the precautions to take when using calcium hypochlorite.

2.2.3 Using Bleach for Water Treatment

Bleach is a readily available and cost-effective option for purifying water. However, it's important to use the correct concentration and follow the proper steps to ensure its effectiveness. Here's a step-by-step guide on using bleach for water treatment:

1. **Choose the right bleach**: Use unscented household bleach that contains 5.25% to 6% sodium hypochlorite as the active ingredient. Avoid using bleach with additives or scents, as they can be harmful.
2. **Filter the water**: If the water is cloudy or contains visible particles, filter it through a clean cloth or coffee filter to remove any debris.
3. **Measure the bleach**: Use a dropper or a clean measuring spoon to add the appropriate amount of bleach to the water. The recommended ratio is 8 drops (or 1/8 teaspoon) of bleach per gallon of water. For

larger quantities, use 1/2 teaspoon of bleach for 5 gallons of water.

4. **Mix thoroughly**: Stir the water vigorously for at least one minute to ensure that the bleach is evenly distributed.
5. **Let it stand**: Allow the water to sit for at least 30 minutes. If the water is very cold or cloudy, let it stand for 60 minutes or longer.
6. **Check for chlorine smell**: After the designated waiting time, the water should have a slight chlorine smell. If it doesn't, repeat the process and double the amount of bleach used.
7. **Aerate the water**: To improve the taste of the treated water, pour it back and forth between clean containers several times to reintroduce oxygen.
8. **Test the water**: If you have water testing strips, you can use them to check the chlorine levels. The water should have a chlorine residual of 0.2 to 4 parts per million (ppm) to ensure its safety for consumption.

Remember, bleach has a shelf life, and its effectiveness diminishes over time. It's important to rotate your supply of bleach regularly and replace it when it reaches its expiration date.

2.2.4 Using Iodine for Water Treatment

Iodine is another effective chemical for water treatment, especially when boiling water is not an option. Here's a step-by-step guide on using iodine for water treatment:

1. **Choose the right iodine**: You can use iodine tablets or tincture for water treatment. Follow the instructions on the packaging for the correct dosage.
2. **Filter the water**: If the water is cloudy or contains visible particles, filter it through a clean cloth or coffee filter to remove any debris.
3. **Add the iodine**: Drop the iodine tablets or the required number of iodine tincture drops into the water. The recommended dosage is typically one tablet or 5 drops of tincture per quart of water.
4. **Mix thoroughly**: Stir the water gently to ensure that the iodine is evenly distributed.
5. **Let it stand**: Allow the water to sit for at least 30 minutes. If the water is very cold or cloudy, let it stand for 60 minutes or longer.
6. **Aerate the water**: Similar to using bleach, pour the treated water back and forth between

clean containers several times to reintroduce oxygen and improve the taste.
7. **Test the water**: If you have water testing strips, you can use them to check the iodine levels. The water should have an iodine residual of 1 to 10 parts per million (ppm) to ensure its safety for consumption.

It's important to note that iodine should not be used by pregnant women, individuals with thyroid problems, or those with iodine allergies. In such cases, alternative water treatment methods should be considered.

2.2.5 Other Chemical Treatment Options

While bleach and iodine are commonly used for chemical water treatment, there are other household items that can be used in emergency situations. These include hydrogen peroxide, vinegar, and even certain types of mouthwash. However, it's important to note that these alternatives may not be as effective or reliable as bleach or iodine. It's always best to use the recommended chemicals and follow the proper guidelines for water treatment.

In this section, we have explored the use of household items for chemical water treatment. By understanding the correct dosage, contact time, and precautions, you can effectively purify water and ensure its safety for consumption. Remember to always have a supply of these chemicals in your emergency preparedness kit, and regularly rotate them to maintain their effectiveness. In the next section, we will delve into the world of filtration systems and explore the different types of water filters available for treating water in crisis situations.

2.3 Filtration Systems

Water filtration systems are an essential tool for preppers to have in their arsenal. When faced with a water crisis, such as contaminated water sources or limited access to clean water, filtration systems can mean the difference between life and death. In this chapter, we will explore the different types of filtration systems available, how to choose the right one for your needs, and how to effectively use them to ensure the water you consume is safe.

2.3.1 Understanding the Basics of Filtration Systems

Before diving into the specifics of different filtration systems, it's important to understand the basic principles behind them. Filtration systems work by removing impurities and contaminants from water, making it safe for consumption. These impurities can include bacteria, viruses, parasites, chemicals, and sediment.

There are two main types of filtration systems: *physical filtration* and **chemical filtration**. Physical filtration involves using a physical barrier, such as a membrane or a mesh, to trap and remove larger particles from the water. Chemical filtration, on the other hand, uses chemical agents, such as activated carbon or iodine, to neutralize or remove contaminants from the water.

2.3.2 Choosing the Right Filtration System

When it comes to choosing the right filtration system for your needs, there are several factors to consider. First and foremost, you need to assess the quality of the water you will be filtering. Is it heavily contaminated with bacteria and viruses, or is it relatively clean but contains some sediment? Understanding the level of contamination will help you determine the type of filtration system you need.

Another important factor to consider is the flow rate of the filtration system. If you need to filter a large volume of water quickly, you will need a system with a higher flow rate. However, keep in mind that higher flow rates may compromise the effectiveness of the filtration system in removing smaller particles and contaminants.

Additionally, consider the portability and ease of use of the filtration system. If you are on the move or in a survival situation, a compact and lightweight system that is easy to assemble and operate will be more practical.

2.3.3 Types of Filtration Systems

There are several types of filtration systems available, each with its own advantages and limitations. Let's explore some of the most common ones:

2.3.3.1 Gravity-Fed Filters

Gravity-fed filters are popular among preppers due to their simplicity and effectiveness. These filters use gravity to force water through a filtration medium, such as ceramic or activated carbon, removing contaminants in the process. They are easy to use and require no electricity or complex setup. However, they may have a slower flow rate compared to other systems.

2.3.3.2 Pump Filters

Pump filters are another popular choice for preppers. These filters use a hand pump to create pressure, forcing water through a filtration medium.

They are generally more efficient than gravity-fed filters and can provide a higher flow rate. However, they require manual effort to operate and may be bulkier and heavier.

2.3.3.3 Straw Filters

Straw filters are compact and lightweight filtration systems that allow you to drink directly from a water source. These filters typically use a combination of physical and chemical filtration to remove contaminants. They are portable and convenient for on-the-go situations but may have a limited lifespan and flow rate.

2.3.3.4 Bottle Filters

Bottle filters are filtration systems built into water bottles or containers. They are designed for personal use and are convenient for filtering water on the move. These filters often use a combination of physical and chemical filtration methods. While they are easy to use, they may have a smaller capacity and slower flow rate compared to other systems.

2.3.4 Using Filtration Systems Effectively

Once you have chosen the right filtration system for your needs, it's important to understand how to use it effectively. Follow these steps to ensure the best results:

- Read the manufacturer's instructions carefully and familiarize yourself with the filtration system.
- If necessary, pre-filter the water to remove larger particles and sediment that may clog the system.
- Assemble the filtration system according to the instructions provided.
- Follow the recommended maintenance and cleaning procedures to ensure the longevity and effectiveness of the system.
- Test the filtered water using a water testing kit to ensure it meets safety standards.

Remember, filtration systems are not foolproof, and they have limitations. Some systems may not effectively remove certain contaminants, such as viruses or chemicals. It's important to understand the capabilities and limitations of your chosen filtration system and have alternative methods of water treatment available if needed.

In the next chapter, we will explore another method of water treatment: solar disinfection. Harnessing the power of the sun, this method can be a valuable addition to your water treatment toolkit. Stay tuned for more essential water survival skills!

2.4 Solar Disinfection

In the remote rural areas of Africa, where clean water is scarce and waterborne diseases run rampant, the introduction of solar disinfection has been nothing short of a game-changer. This innovative method harnesses the power of the sun to purify water, bringing about a dramatic reduction in illness and a significant improvement in the overall health of these communities. It serves as a powerful reminder of the vital role water plays as a resource, particularly in times of crisis, and emphasizes the importance of being prepared for water-related emergencies.

This chapter delves into the world of solar disinfection, showcasing its effectiveness and simplicity as a key technique in combating waterborne illnesses. It stands as a testament to the incredible impact sustainable solutions can have on water quality, highlighting the adaptability and resilience needed to ensure access to clean, safe drinking water. Solar disinfection emerges as an essential strategy for both preppers and vulnerable communities, demonstrating the power of proactive measures in safeguarding public health and well-being.

2.4.1 Harnessing the Power of the Sun

When disaster strikes and clean water becomes scarce, it is essential to have alternative methods of water treatment. Solar disinfection, also known as SODIS, is a technique that utilizes the sun's ultraviolet (UV) radiation to kill harmful microorganisms in water. This method is particularly useful in situations where other treatment options are unavailable or impractical.

2.4.2 How Solar Disinfection Works

Solar disinfection works by utilizing two natural processes: heat and UV radiation. When water is exposed to sunlight, the heat from the sun raises the temperature, while the UV radiation kills the microorganisms present in the water. This combination effectively disinfects the water, making it safe for consumption.

2.4.3 Implementing Solar Disinfection

To implement solar disinfection, follow these simple steps:

1. **Select a clear plastic or glass container**: Choose a container that is transparent and has a tight-fitting lid. This will allow sunlight to

penetrate the water while preventing contamination.
2. **Fill the container with water**: Fill the container with clear water from a reliable source. It is essential to use water that is free from visible particles or debris.
3. **Place the container in direct sunlight**: Find a sunny spot outdoors and place the container in direct sunlight. It is recommended to leave the container undisturbed for at least six hours on a sunny day.
4. **Let the sun do its work**: The heat and UV radiation from the sun will work together to disinfect the water. During this time, the microorganisms present in the water will be destroyed, making it safe to drink.
5. **Store the treated water**: Once the disinfection process is complete, carefully store the treated water in clean, covered containers. It is crucial to protect the water from recontamination.

2.4.4 Factors Affecting Solar Disinfection

While solar disinfection is a simple and effective method, several factors can affect its efficiency:

- **Climate**: Solar disinfection works best in sunny climates with high UV radiation.

Cloudy or overcast days may require longer exposure times.
- **Water quality**: The clarity of the water plays a significant role in the effectiveness of solar disinfection. Turbid or cloudy water may require additional filtration or settling before treatment.
- **Container material**: The container used for solar disinfection should be transparent and made of clear plastic or glass. Avoid using containers made of materials that may leach harmful substances into the water.
- **Altitude**: Higher altitudes may require longer exposure times due to the reduced intensity of UV radiation.

2.4.5 Advantages and Limitations of Solar Disinfection

Solar disinfection offers several **advantages** that make it a valuable technique for preppers:

- **Cost-effective**: Solar disinfection requires minimal equipment and relies solely on the power of the sun, making it a cost-effective method of water treatment.
- **Simple and accessible**: The process of solar disinfection is easy to understand and

implement, making it accessible to individuals with limited resources or technical knowledge.
- **Portable and scalable**: Solar disinfection can be used on a small scale, treating water for individual use, or on a larger scale, treating water for communities.

However, it is important to note the **limitations** of solar disinfection:

- **Time-consuming**: Solar disinfection requires several hours of exposure to sunlight, which may not be practical in emergency situations where immediate access to clean water is crucial.
- **Climate-dependent**: The effectiveness of solar disinfection is highly dependent on the availability of sunlight and UV radiation. Cloudy or overcast weather can significantly impact the disinfection process.
- **Limited to microbial contaminants**: Solar disinfection is effective in killing most microorganisms present in water but does not remove chemical contaminants or heavy metals.

2.4.7 Summary

Solar disinfection is a valuable technique for preppers to treat water in emergency situations. By harnessing the power of the sun, this method provides a simple and cost-effective way to disinfect water and make it safe for consumption. While it has its limitations, solar disinfection can be a lifesaver when other treatment options are unavailable. Remember to consider factors such as climate, water quality, and container material when implementing solar disinfection.

Rainwater Harvesting and Collection

3.1 The Basics of Rainwater Harvesting

Rainwater harvesting is a crucial skill for preppers to master in order to ensure a reliable and sustainable water supply during times of crisis. When disaster strikes and traditional water sources become compromised or inaccessible, having the knowledge and tools to collect and utilize rainwater can mean the difference between survival and desperation.

3.1.1 Why Rainwater Harvesting Matters

Water is the essence of life. It sustains us, nourishes us, and keeps our bodies functioning properly. In a survival situation, access to clean and safe water becomes even more critical. Without it, dehydration, illness, and even death can quickly become a reality.

Rainwater harvesting offers a practical and effective solution to this problem. By capturing and storing rainwater, preppers can create a reliable source of water that is not dependent on external infrastructure or resources. This independence is invaluable during times of crisis when municipal water systems may be compromised or contaminated.

3.1.2 The Rainwater Harvesting Process

Rainwater harvesting involves the collection, storage, and treatment of rainwater for various uses. The process begins with the collection of rainwater from rooftops, gutters, or other surfaces. This water is then directed into storage containers, where it can be treated and used for drinking, cooking, hygiene, and other essential needs.

3.1.3 Collecting Rainwater Safely

When collecting rainwater, it is important to ensure that the water is free from contaminants and safe for consumption. This can be achieved by using clean and properly maintained collection surfaces, such as roofs or specially designed catchment systems. It is also crucial to regularly clean and inspect these surfaces to prevent the buildup of debris or pollutants.

3.1.4 Storing Rainwater

Proper storage of rainwater is essential to maintain its quality and prevent the growth of harmful bacteria or algae. Rainwater storage containers should be made of food-grade materials, such as high-density polyethylene (HDPE) or stainless steel, to ensure that they do not leach any harmful substances into the water. Additionally, the containers should be tightly sealed to prevent the entry of insects, rodents, or other contaminants.

3.1.5 Treating Rainwater for Consumption

While rainwater is generally considered safe for non-potable uses, such as gardening or cleaning, it may require treatment before it can be consumed. Common methods of rainwater treatment include filtration, disinfection, and chemical treatment. Filtration systems can remove larger particles and sediment, while disinfection methods, such as boiling or using chlorine tablets, can kill harmful bacteria and viruses. It is important to follow proper treatment protocols and guidelines to ensure the safety of the harvested rainwater.

3.1.6 Maximizing Rainwater Harvesting Efficiency

To maximize the efficiency of rainwater harvesting, it is important to consider factors such as rainfall patterns, collection surface area, and storage capacity. Understanding the local climate and rainfall patterns can help determine the optimal size of the collection system and storage containers. Additionally, implementing strategies such as rainwater diversion, gutter maintenance, and using rain barrels or tanks with larger capacities can help capture and store more rainwater during periods of heavy rainfall.

3.2 DIY Rainwater Collection Systems

Rainwater is a valuable resource that can be harnessed and utilized in times of crisis. In this chapter, we will explore the world of DIY rainwater collection systems and learn how to build our own setup. By doing so, we can ensure a sustainable and reliable source of water for our survival needs.

3.2.1 The Basics of Rainwater Harvesting

Before we dive into the world of DIY rainwater collection systems, it's important to understand the basics of rainwater harvesting. Rainwater harvesting is the process of collecting and storing rainwater for later use. It is a simple and effective way to make the most of this precious resource.

In this section, we will explore the different components of a rainwater collection system and their functions. We will discuss the importance of gutters, downspouts, and filters in capturing and directing rainwater into storage containers. We will also delve into the various types of storage containers available and their pros and cons.

3.2.2 Building Your Own Rainwater Collection System

Now that we understand the basics of rainwater harvesting, it's time to roll up our sleeves and build our own DIY rainwater collection system. Building a rainwater collection system can be a fun and rewarding project that allows us to take control of our water supply.

In this section, we will provide step-by-step instructions on how to build a simple rainwater collection system. We will discuss the materials needed, such as gutters, downspouts, storage containers, and filters. We will guide you through the process of installing the gutters and downspouts, connecting them to the storage containers, and setting up the filtration system.

3.2.3 Maintaining and Upgrading Your Rainwater Collection System

Once you have built your DIY rainwater collection system, it's important to maintain and periodically upgrade it to ensure its efficiency and longevity. In this section, we will discuss the maintenance tasks required to keep your system in optimal condition.

We will cover topics such as cleaning the gutters and filters, inspecting the storage containers for leaks or damage, and ensuring proper water flow through the system. We will also provide tips on how to upgrade your system to increase its capacity or improve its functionality.

In conclusion, building a DIY rainwater collection system is a practical and empowering way to secure a sustainable water source. By understanding the basics of rainwater harvesting, following step-by-step instructions, and maintaining and upgrading our system, we can be well-prepared for water-related emergencies. So, let's grab our tools and get started on building our own rainwater collection system!

3.3 Rainwater Purification

Rainwater is a valuable resource that can be harvested and used for various purposes, especially in times of crisis or water scarcity. However, it is important to remember that rainwater is not always safe for consumption without proper purification. In this chapter, we will explore the essential techniques and methods for purifying rainwater to ensure its safety for drinking and other uses.

3.3.1 Understanding Rainwater Contamination

Before we delve into the purification methods, it is crucial to understand the potential contaminants that can be present in rainwater. Rainwater can pick up pollutants and contaminants from various sources, including air pollution, bird droppings, dust, and debris on rooftops or collection surfaces. These contaminants can pose health risks if consumed without proper treatment.

To ensure the safety of rainwater, it is essential to implement effective purification techniques that remove or neutralize these contaminants. Let's explore some of the most reliable methods for rainwater purification.

3.3.2 Boiling Rainwater

Boiling water is one of the oldest and most reliable methods of purification. By bringing rainwater to a rolling boil for at least one minute (or three minutes at higher altitudes), you can effectively kill most microorganisms and pathogens present in the water. Boiling is a simple and accessible method that requires minimal equipment, making it an excellent option for emergency situations.

3.3.3 Chemical Treatment

Chemical treatment is another effective method for purifying rainwater. Household items such as chlorine bleach or iodine tablets can be used to disinfect the water and kill harmful bacteria, viruses, and parasites. It is important to follow the instructions provided with the chemical treatment products and use the correct dosage to ensure effective purification.

3.3.4 Filtration Systems

Filtration systems are an essential tool for rainwater purification, as they can effectively remove physical impurities and some microorganisms. There are various types of water filters available, including activated carbon filters, ceramic filters, and reverse osmosis filters. Each type has its own advantages and limitations, so it is important to choose the right filter based on your specific needs and the level of purification required.

3.3.5 Ultraviolet (UV) Treatment

UV treatment is a modern and efficient method for purifying rainwater. UV light can effectively destroy the DNA of microorganisms, rendering them unable to reproduce and cause harm. UV treatment devices are compact and easy to use, making them a convenient option for purifying rainwater on the go. However, it is important to ensure that the UV device is properly maintained and the water is exposed to the UV light for the recommended duration to achieve effective purification.

3.3.6 Combination Methods

In some cases, combining multiple purification methods can provide an extra layer of safety and ensure comprehensive purification of rainwater. For example, you can start by filtering the water to remove physical impurities, then use chemical treatment or UV treatment to kill any remaining microorganisms. By using a combination of methods, you can enhance the effectiveness of rainwater purification and ensure the safety of the water for consumption.

3.3.7 Rainwater Purification Tips and Best Practices

To further assist you in purifying rainwater effectively, here are some tips and best practices to keep in mind:

- Always collect rainwater from clean surfaces, such as rooftops or specially designed

collection systems, to minimize contamination.
- Regularly clean and maintain your rainwater collection system to prevent the buildup of debris and contaminants.
- Use food-grade containers for storing rainwater to avoid leaching of harmful chemicals into the water.
- Test the quality of your rainwater periodically to ensure that your purification methods are effective and the water is safe for consumption.
- Educate yourself and your family members about the importance of rainwater purification and the proper techniques to follow.

Remember, rainwater can be a valuable resource during times of crisis, but it must be properly purified before use. By implementing the techniques and best practices outlined in this chapter, you can ensure the safety and reliability of your rainwater supply for drinking and other essential needs. Stay prepared, stay safe!

3.4 Creative Uses of Rainwater

Rainwater is a precious resource that can be utilized in various ways beyond just drinking and cooking. In a crisis situation, where water scarcity is a constant concern, preppers need to think outside the box and explore creative uses for rainwater. By harnessing the power of rain, you can maximize its potential and ensure your survival in more ways than one.

3.4.1 Watering Your Survival Garden

One of the most practical uses of rainwater is for watering your survival garden. In times of crisis, when access to clean water is limited, using rainwater to nourish your plants can help ensure a sustainable food source. Rainwater is naturally free of chemicals and additives, making it an ideal choice for irrigation. By collecting rainwater in barrels or tanks, you can create a reliable and eco-friendly watering system for your garden.

3.4.2 Cleaning and Laundry

In a water-scarce situation, maintaining personal hygiene and cleanliness becomes a challenge. However, rainwater can be a valuable resource for cleaning purposes. You can use rainwater to wash your hands, bathe, and even do laundry. By setting up a simple filtration system, you can remove any impurities from the rainwater and use it for these essential tasks. This not only conserves your precious drinking water but also ensures that you can maintain good hygiene even in challenging circumstances.

3.4.3 Household Cleaning

Rainwater can also be used for various household cleaning tasks. From mopping the floors to washing dishes, rainwater can be a valuable substitute for tap water. By using rainwater for these purposes, you can conserve your drinking water supply and ensure that it is reserved for essential needs. Additionally, rainwater is often softer than tap water, which can make cleaning more effective and efficient.

3.4.4 Watering Livestock

If you have livestock or animals that rely on a water supply, rainwater can be a lifeline. By collecting rainwater in large containers or troughs, you can provide your animals with a clean and reliable water source. This not only ensures their survival but also reduces the strain on your drinking water supply. Remember to filter the rainwater before giving it to your animals to ensure their health and well-being.

3.4.5 Construction and DIY Projects

Rainwater can also be used for construction and DIY projects. Whether you're building a shelter, repairing structures, or working on other projects, rainwater can be a valuable resource. It can be used for mixing cement, cleaning tools, and even as a substitute for tap water in various construction tasks. By utilizing rainwater for these purposes, you can conserve your drinking water supply and ensure that it is reserved for essential needs.

3.4.6 Fire Protection

In a crisis situation, the risk of fires can increase due to various factors. Having a reliable source of water for fire protection is crucial for your safety and the preservation of your belongings. Rainwater can be stored in large containers or tanks and used for firefighting purposes. By strategically placing these containers around your property, you can have quick access to water in case of an emergency. Remember to have appropriate firefighting equipment and knowledge to handle such situations safely.

3.4.7 Bartering and Trade

In a post-apocalyptic world, where resources are scarce, rainwater can become a valuable commodity for bartering and trade. As clean water becomes increasingly rare, those who have access to rainwater can use it as a means of exchange for other essential goods or services. By collecting and storing rainwater, you not only ensure your own survival but also have a valuable asset that can be traded for other necessities.

3.4.8 Emotional and Psychological Support

While not a tangible use, rainwater can also provide emotional and psychological support during a crisis. The sound of rain falling and the sight of water replenishing the earth can have a calming and soothing effect on the mind. By connecting with nature and appreciating the beauty of rainwater, you can find solace and strength in challenging times. Take a moment to pause, listen to the rain, and let it remind you of the resilience and adaptability of the human spirit.

In conclusion, rainwater is a versatile resource that can be utilized in various creative ways during a crisis. From watering your survival garden to providing for your livestock, rainwater can be a lifeline in times of water scarcity. By exploring these creative uses of rainwater, you can maximize its potential and ensure your survival in more ways than one. Remember to always collect and store rainwater safely, and make the most of this precious resource.

3.5 Legal Considerations

When it comes to rainwater harvesting, it's not just about setting up your collection system and purifying the water. There are also legal considerations that you need to be aware of. While rainwater is a free and abundant resource, it is important to understand the regulations and restrictions that may apply to its collection and use. In this section, we will explore the legal aspects of rainwater harvesting and provide you with the information you need to navigate these considerations.

3.5.1 Understanding Rainwater Harvesting Regulations

Before you start collecting rainwater, it is crucial to familiarize yourself with the local regulations and laws regarding rainwater harvesting. These regulations can vary from state to state and even within different municipalities. Some areas may have specific guidelines and permits that need to be obtained before you can legally collect rainwater.

To begin, research the laws and regulations in your area. Check with your local government or water authority to find out if there are any restrictions or requirements for rainwater harvesting. They can provide you with the necessary information and guide you through the process.

In some cases, rainwater harvesting may be completely unrestricted, allowing you to collect and use rainwater as you see fit. However, in other areas, there may be limitations on the amount of rainwater you can collect or restrictions on its use. For example, some regions may prohibit the use of rainwater for potable purposes, while others may require a permit for large-scale rainwater collection systems.

3.5.2 Obtaining Permits and Approvals

If your local regulations require permits or approvals for rainwater harvesting, it is essential to follow the necessary procedures. This typically involves submitting an application to the appropriate authority and providing details about your rainwater collection system, including its design, capacity, and intended use.

The permitting process may also involve inspections to ensure that your system meets safety and quality standards. These inspections are important to ensure that your rainwater collection system is properly installed and does not pose any risks to public health or the environment.

While the permitting process may seem like a hassle, it is crucial to comply with the regulations in your area. Failure to do so can result in fines or other legal consequences. By obtaining the necessary permits and approvals, you can ensure that your rainwater harvesting activities are legal and in compliance with local regulations.

3.5.3 Water Rights and Ownership

Another important legal consideration when it comes to rainwater harvesting is water rights and ownership. In some areas, water rights are governed by complex legal frameworks that determine who has the right to use and access water resources. These frameworks may include laws related to riparian rights, prior appropriation, or other legal doctrines.

It is important to understand the water rights laws in your area to ensure that you are not infringing on the rights of others or violating any legal obligations. In some cases, you may need to obtain water rights or permissions from the appropriate authorities before you can legally collect and use rainwater.

Additionally, it is important to consider the ownership of rainwater once it is collected. In some jurisdictions, rainwater may be considered a common resource that cannot be owned by individuals. However, in other areas, rainwater may be subject to private ownership or shared ownership with the government or other entities.

Understanding the legal aspects of water rights and ownership will help you navigate any potential conflicts or issues that may arise from rainwater harvesting. It is always advisable to consult with legal professionals or experts in water law to ensure that you are in compliance with the applicable regulations.

3.5.4 Liability and Insurance

When setting up a rainwater harvesting system, it is important to consider liability and insurance implications. While rainwater is generally safe for non-potable uses, there is always a risk of contamination or other issues that could lead to health or property damage.

Before implementing a rainwater collection system, it is advisable to consult with your insurance provider to understand any coverage limitations or requirements. They can provide guidance on whether your system is covered under your existing policy or if you need to obtain additional coverage.

Liability is another important consideration. If you are using rainwater for non-potable purposes, such as irrigation or toilet flushing, the risk of liability may be lower. However, if you are using rainwater for potable purposes or distributing it to others, there may be additional legal responsibilities and potential liabilities.

To protect yourself from liability, it is advisable to follow best practices for rainwater harvesting, including proper system design, maintenance, and water treatment. Additionally, consider consulting with legal professionals to understand any specific liability issues that may apply in your jurisdiction.

3.5.5 Conclusion

While rainwater harvesting is a valuable and sustainable practice, it is essential to understand the legal considerations that come with it. By familiarizing yourself with the regulations, obtaining necessary permits, understanding water rights, and considering liability and insurance implications, you can ensure that your rainwater harvesting activities are legal, safe, and compliant with the law. Remember, compliance with the law not only protects you from legal consequences but also helps to promote responsible and sustainable water use in your community.

Water Conservation and Efficiency

4.1 The Importance of Water Conservation in Crisis

Water is the essence of life. It sustains us, nourishes us, and keeps us alive. In a crisis situation, access to clean and safe water becomes even more critical. As a prepper, you understand the importance of being prepared for any emergency, and that includes having a plan for water conservation. In this chapter, we will explore why water conservation is crucial during a crisis and provide you with practical tips to reduce water waste and ensure a sustainable water supply.

4.1.1 The Value of Water in Survival

Imagine a world without water. It's a terrifying thought, isn't it? Water is not only essential for drinking and cooking, but it also plays a vital role in personal hygiene, sanitation, and food production. Without an adequate supply of water, our bodies become dehydrated, our crops wither, and our hygiene and sanitation practices suffer, leading to the spread of diseases. In a crisis situation, where water sources may be contaminated or scarce, the value of water becomes even more apparent.

4.1.2 Understanding Water Scarcity

Water scarcity is a growing concern worldwide, and it can be exacerbated during a crisis. Natural disasters, such as droughts or floods, can disrupt water supplies, leaving communities without access to clean water. Additionally, infrastructure failures or contamination incidents can render municipal water systems unsafe to use. By understanding the potential for water scarcity in crisis situations, you can better prepare and conserve this precious resource.

4.1.3 Reducing Water Waste: Practical Tips for Everyday Life

Conserving water is not just a practice for emergencies; it should be a part of our everyday lives. By adopting water-saving habits, we can reduce our water consumption and contribute to a more sustainable future. In this section, we will explore practical tips for conserving water in your daily activities. From fixing leaky faucets and using efficient appliances to taking shorter showers and collecting rainwater, every small action can make a significant difference.

4.1.4 Greywater Recycling: Reusing Water for Non-Potable Purposes

Greywater recycling is a technique that allows you to reuse water from sources such as sinks, showers, and washing machines for non-potable purposes. By implementing greywater recycling systems, you can reduce your reliance on freshwater sources and conserve water for essential needs. We will discuss the benefits of greywater recycling, different methods of treatment, and how to safely use recycled water in your home or garden.

4.1.5 Xeriscaping: Landscaping Techniques for Water Efficiency

Traditional landscaping practices often require excessive amounts of water to maintain lush lawns and gardens. Xeriscaping, on the other hand, is a landscaping technique that focuses on water efficiency and conservation. By choosing drought-tolerant plants, implementing efficient irrigation systems, and using mulch to retain moisture, you can create a beautiful and sustainable landscape that requires minimal water. We will provide you with practical tips and design ideas to help you transform your outdoor space into a water-efficient oasis.

4.1.6 Conclusion

Water conservation is not just a responsibility; it is a necessity. By understanding the importance of water conservation in crisis situations and implementing practical strategies, you can ensure a sustainable water supply for yourself and your loved ones. So let's dive into the world of water conservation and learn how to make every drop count.

4.2 Reducing Water Waste

Water is a precious resource, especially in times of crisis. As a prepper, it is crucial to not only have access to clean and safe water but also to use it wisely. In this chapter, we will explore practical tips and strategies to help you reduce water waste in your everyday life. By implementing these techniques, you can stretch your water supply further and increase your chances of long-term survival.

4.2.1 Fix Those Leaks!

One of the most common sources of water waste in households is leaky faucets and pipes. Even a small drip can add up to gallons of wasted water over time. It is essential to regularly check for leaks and fix them promptly. Here are some steps you can take:

- Check all faucets, showerheads, and toilets for leaks. A simple way to detect leaks is by placing a few drops of food coloring in the toilet tank. If the color appears in the bowl without flushing, you have a leak.
- Replace worn-out washers or faulty parts in faucets and showerheads.
- Use plumber's tape or sealant to fix any leaks in pipes.
- Consider installing low-flow faucets and showerheads to reduce water usage.

4.2.2 Be Mindful of Your Water Habits

Being mindful of your water habits can go a long way in reducing water waste. Here are some practical tips to help you conserve water:

- Turn off the faucet while brushing your teeth or shaving. This simple habit can save gallons of water each day.
- Take shorter showers. Aim to keep your showers under five minutes to minimize water usage.
- Collect the cold water that runs before the hot water arrives and use it for watering plants or flushing toilets.

- Only run the dishwasher and washing machine with full loads to maximize water efficiency.
- Use a basin or plug the sink when washing dishes. This way, you can reuse the water for multiple dishes.
- Use a broom instead of a hose to clean outdoor areas like driveways and patios.
- Opt for a bucket and sponge when washing your car instead of using a hose.

4.2.3 Upgrade Your Appliances

Older appliances, such as toilets, washing machines, and dishwashers, can be major water wasters. Consider upgrading to more water-efficient models to reduce your water consumption. Look for appliances with the WaterSense label, which indicates that they meet water efficiency standards set by the Environmental Protection Agency (EPA).

4.2.4 Harvest and Reuse Water

Another effective way to reduce water waste is by harvesting and reusing water. Here are some methods you can implement:

- Install rain barrels to collect rainwater from your roof. This water can be used for watering plants or flushing toilets.

- Use a greywater system to collect and treat water from sinks, showers, and washing machines. This treated water can be used for irrigation or flushing toilets.
- Consider installing a composting toilet, which uses little to no water for waste disposal.

4.2.5 Educate and Involve Your Household

Reducing water waste is a collective effort that involves everyone in your household. Educate your family members about the importance of water conservation and involve them in implementing water-saving practices. Here are some ways to get everyone on board:

- Have a family meeting to discuss the importance of water conservation and the specific actions you will take as a household.
- Assign responsibilities to each family member, such as checking for leaks, monitoring water usage, or implementing water-saving habits.
- Create a reward system to encourage everyone's participation and celebrate achievements in water conservation.

By reducing water waste in your everyday life, you not only conserve this precious resource but also increase your self-sufficiency in times of crisis.

Implement the tips and strategies discussed in this chapter to make the most of your water supply and ensure your long-term survival.

Remember, every drop counts!

4.3 Greywater Recycling

Water is a precious resource, especially in times of crisis. As a prepper, it is crucial to be prepared for water-related emergencies and have the knowledge and skills to handle them effectively. In this chapter, we will explore the concept of greywater recycling, a technique that allows you to reuse water for non-potable purposes. By implementing greywater recycling systems, you can significantly reduce your water consumption and increase your self-sufficiency during challenging times.

4.3.1 Understanding Greywater

Greywater refers to the wastewater generated from various household activities, excluding toilet waste. It includes water from sinks, showers, baths, and washing machines. While greywater is not suitable for drinking, it can be safely used for other purposes such as irrigation, flushing toilets, and cleaning. By recycling greywater, you can conserve freshwater resources and reduce the strain on your water supply.

4.3.2 Greywater Recycling Systems

There are several types of greywater recycling systems that you can implement in your home. The simplest and most common system is the "bucket and dipper" method, where you manually collect greywater in buckets and use it for specific tasks. This method is low-cost and easy to set up, making it a great option for temporary or emergency situations.

Another popular system is the "branched drain" system, which involves diverting greywater from your sinks, showers, and washing machines to a separate drainage system. This system requires some plumbing work but can be more efficient and convenient in the long run.

For those looking for a more advanced solution, there are greywater treatment systems available that filter and purify the greywater for reuse. These systems use various filtration methods such as sand filters, biological filters, and disinfection processes to ensure the water is safe for non-potable purposes.

4.3.3 Benefits of Greywater Recycling

Implementing greywater recycling systems in your home offers numerous benefits. Firstly, it helps conserve freshwater resources by reducing the amount of water you need from external sources. This is particularly important during water scarcity or when access to clean water is limited.

Secondly, greywater recycling can save you money on your water bills. By reusing water for tasks like irrigation or toilet flushing, you can significantly reduce your overall water consumption, leading to lower utility costs.

Furthermore, greywater recycling promotes self-sufficiency and resilience. During a crisis, when water supplies may be disrupted, having a reliable source of non-potable water can make a significant difference in your ability to meet your basic needs.

4.3.4 Greywater Recycling Tips and Considerations

When implementing a greywater recycling system, there are a few important tips and considerations to keep in mind:

- **Safety First**: It is crucial to ensure that the greywater you are recycling does not come into contact with your drinking water supply or pose any health risks. Avoid using greywater for tasks that involve direct contact with food or areas where cross-contamination can occur.

- **Proper Filtration**: If you are using greywater for irrigation, make sure to filter it properly to remove any solid particles or contaminants. This can be done through simple filtration methods such as using mesh screens or fabric filters.
- **Avoid Chemical Contamination**: Be mindful of the products you use in your household that may contaminate the greywater. Avoid using harsh chemicals or toxic substances that can harm plants or the environment.
- **Regular Maintenance**: Regularly inspect and maintain your greywater recycling system to ensure its proper functioning. Clean filters, check for leaks, and address any issues promptly to avoid potential problems.
- **Educate and Communicate**: If you live with others, make sure everyone in your household understands the greywater recycling system and knows how to use it correctly. Clear communication and education are essential to

ensure the system is used effectively and safely.

Summary

Greywater recycling is a valuable technique that allows you to reuse water for non-potable purposes, reducing your water consumption and increasing your self-sufficiency. By understanding the different greywater recycling systems, considering the benefits, and following the necessary tips and considerations, you can implement an effective greywater recycling system in your home. Remember, every drop counts, and by recycling greywater, you contribute to water conservation and resilience in times of crisis.

4.4 Xeriscaping

In times of crisis, water becomes a precious resource that must be conserved and used wisely. As a prepper, it is essential to be prepared for water-related emergencies and have the knowledge and skills to handle them effectively. In this chapter, we will explore the concept of xeriscaping, a landscaping technique that promotes water efficiency and conservation. By implementing xeriscaping principles, you can create a beautiful and sustainable outdoor space while minimizing your water usage.

4.4.1 What is Xeriscaping?

Xeriscaping is a landscaping approach that focuses on designing and maintaining gardens and landscapes with minimal water requirements. The term "xeriscape" originates from the Greek word "xeros," meaning dry, and "scape," referring to a view or scene. Xeriscaping aims to create visually appealing landscapes that are well-suited to the local climate and require little to no supplemental irrigation.

4.4.2 The Benefits of Xeriscaping

Xeriscaping offers numerous benefits, making it an ideal choice for preppers looking to conserve water and create a sustainable outdoor environment. Here are some of the key advantages of xeriscaping:

- **Water Conservation**: Xeriscaping significantly reduces water usage compared to traditional landscaping methods. By selecting drought-tolerant plants, implementing efficient irrigation systems, and using mulch to retain moisture, you can conserve water and minimize waste.
- **Cost Savings**: With xeriscaping, you can reduce your water bills and maintenance costs. By relying on native plants that are adapted to

the local climate, you can eliminate the need for excessive watering, fertilizers, and pesticides.

- **Environmental Sustainability**: Xeriscaping promotes environmental sustainability by reducing the strain on water resources. By conserving water, you contribute to the preservation of natural habitats and ecosystems.
- **Increased Property Value**: A well-designed xeriscape can enhance the aesthetic appeal of your property and increase its value. Potential buyers appreciate low-maintenance landscapes that require minimal water and upkeep.

4.4.3 Designing a Xeriscape

Creating a xeriscape involves careful planning and consideration of various factors, including climate, soil conditions, and plant selection. Here are some key steps to follow when designing your xeriscape:

- **Assess Your Site**: Evaluate the existing conditions of your landscape, including sun exposure, soil type, and drainage. This information will help you determine the appropriate plants and design elements for your xeriscape.

- **Choose Drought-Tolerant Plants**: Select plants that are well-adapted to your local climate and require minimal watering. Native plants are often the best choice as they are naturally suited to the region's conditions.
- **Group Plants by Water Needs**: Arrange your plants in zones based on their water requirements. This allows for more efficient irrigation, as plants with similar needs can be watered together.
- **Implement Efficient Irrigation**: Install drip irrigation systems or soaker hoses to deliver water directly to the plant's root zone. This minimizes water loss through evaporation and ensures that plants receive the necessary moisture.
- **Mulch and Soil Improvement**: Apply a layer of organic mulch around plants to retain moisture, suppress weed growth, and regulate soil temperature. Additionally, amend the soil with organic matter to improve its water-holding capacity.
- **Consider Hardscape Elements**: Incorporate hardscape features such as pathways, patios, and rock gardens into your xeriscape design. These elements not only add

visual interest but also reduce the amount of water-intensive turfgrass.

4.4.4 Maintaining a Xeriscape

Once your xeriscape is established, proper maintenance is crucial to ensure its long-term success. Here are some maintenance practices to keep in mind:

- **Water Efficiently**: Monitor your plants' water needs and adjust irrigation accordingly. Avoid overwatering, as it can lead to root rot and other plant diseases. Regularly check for leaks or malfunctions in your irrigation system.
- **Weed Control**: Keep your xeriscape weed-free to prevent competition for water and nutrients. Apply mulch regularly to suppress weed growth, and manually remove any weeds that do appear.
- **Pruning and Trimming**: Regularly prune and trim your plants to maintain their shape and promote healthy growth. Remove dead or diseased branches to prevent the spread of pests and diseases.
- **Soil Maintenance**: Periodically test your soil's pH and nutrient levels to ensure optimal plant health. Amend the soil as needed with organic matter or appropriate fertilizers.

- **Seasonal Adjustments**: Adjust your xeriscape maintenance practices based on seasonal changes. During periods of drought or extreme heat, you may need to provide additional water to ensure plant survival.

By implementing xeriscaping techniques, you can create a beautiful and sustainable landscape that conserves water and thrives in times of crisis. Remember to choose drought-tolerant plants, design efficient irrigation systems, and maintain your xeriscape regularly. With these practices in place, you can enjoy a resilient and water-efficient outdoor space that complements your prepping efforts.

Emergency Water Sources

5.1 Natural Water Sources

Water is the essence of life, and in a crisis situation, access to clean and safe water becomes even more crucial. As a prepper, it is essential to be prepared for any water-related emergency, and that means understanding where to find water when traditional sources are unavailable. In this chapter, we will explore natural water sources such as rivers, lakes, and streams, and learn how to safely utilize them for survival.

5.1.1 Rivers: The Lifelines of Nature

Rivers are nature's lifelines, flowing through vast landscapes and providing a constant supply of freshwater. When disaster strikes, rivers can become a valuable source of water for survival. However, it is important to exercise caution when using river water, as it may contain contaminants and pose health risks.

To safely utilize river water, it is crucial to understand the potential hazards and take appropriate measures to treat it. In this section, we will discuss the importance of identifying upstream pollution sources, such as industrial areas or agricultural runoff, which can contaminate river water. We will also explore various water treatment techniques, including filtration and chemical disinfection, to ensure that the water is safe for consumption.

5.1.2 Lakes: Nature's Reservoirs

Lakes are another natural water source that can provide a lifeline during a crisis. They are often larger and more stable than rivers, making them a potentially more reliable source of water. However, it is important to remember that not all lakes are safe for consumption.

In this section, we will discuss the importance of assessing the quality of lake water before using it for drinking or cooking. We will explore the signs of contamination, such as algae blooms or foul odors, and provide guidance on how to treat lake water to make it safe for consumption. Additionally, we will discuss the importance of understanding the local ecosystem and its impact on lake water quality.

5.1.3 Streams: Nature's Hidden Gems

Streams are often overlooked as a water source, but they can be a valuable resource during a crisis. These smaller bodies of flowing water can be found in various terrains, including forests and mountains, and can provide a reliable source of freshwater.

In this section, we will explore the benefits and challenges of using stream water for survival. We will discuss the importance of locating streams in safe and accessible areas, away from potential pollution sources. Additionally, we will delve into the various methods of collecting and treating stream water, such as using a portable water filter or employing natural filtration techniques.

5.1.4 Natural Water Sources: A Summary

In this section, we have explored the importance of natural water sources such as rivers, lakes, and streams in a survival situation. We have discussed the potential hazards and contaminants that may be present in these water sources and provided guidance on how to safely collect and treat the water for consumption.

Remember, when utilizing natural water sources, always prioritize your safety and take necessary precautions to ensure the water is free from contaminants. By understanding the characteristics of each water source and employing appropriate treatment techniques, you can ensure a reliable supply of clean and safe water in times of crisis.

5.2 Urban Water Sources

In times of crisis, when water scarcity becomes a pressing issue, preppers must be resourceful and adaptable. While natural water sources such as rivers and lakes are often the go-to options for survival, urban environments present a unique set of challenges and opportunities when it comes to accessing water. In this section, we will explore the various urban water sources that preppers can tap into to ensure their water needs are met during emergencies.

5.2.1 Municipal Water Systems: A Lifeline in the Concrete Jungle

When disaster strikes, municipal water systems can be a lifeline for urban preppers. These systems are designed to provide clean and safe water to the population, and understanding how they work can give you an advantage in securing water during a crisis.

Municipal water systems typically consist of water treatment plants, distribution networks, and storage facilities. Water is sourced from natural sources such as rivers, lakes, or underground aquifers. It is then treated to remove impurities and distributed through a network of pipes to homes and businesses.

To tap into the municipal water system during an emergency, it is crucial to know the location of the nearest water treatment plant and distribution center. Familiarize yourself with the layout of the water distribution network in your area, including the location of valves and hydrants. This knowledge will enable you to access water directly from the system if your regular water supply is disrupted.

During a crisis, it is possible that the municipal water system may become compromised or shut down. In such situations, it is essential to have alternative methods of accessing water. This could include storing water in advance, utilizing rainwater harvesting systems, or exploring other unconventional water sources.

5.2.2 Unconventional Water Sources: Ponds, Wells, and Cisterns

In addition to the municipal water system, urban preppers can also consider tapping into unconventional water sources to meet their water needs. These sources include ponds, wells, and cisterns, which can provide a reliable and independent water supply during emergencies.

Ponds are often found in urban parks or private properties and can serve as a valuable water source.

However, it is crucial to ensure the water is safe for consumption by treating it properly. Boiling, filtration, or chemical treatment methods can be employed to purify the water and make it suitable for drinking.

Wells are another potential source of water in urban areas. While many urban dwellers may not have access to private wells, some older buildings or neighborhoods may still have functioning wells. It is important to test the water quality regularly and have the necessary equipment to extract water from the well.

Cisterns, commonly found in older buildings, are large storage tanks that collect rainwater from rooftops. These tanks can hold a significant amount of water, making them an excellent resource during water shortages. However, it is crucial to ensure the cistern is properly maintained and the water is treated before consumption.

When utilizing unconventional water sources, it is essential to have the necessary equipment and knowledge to treat and store the water properly. Understanding the potential risks and implementing appropriate treatment methods will ensure a safe and reliable water supply.

5.2.3 Harvesting Water from Urban Landscapes

Urban environments offer unique opportunities for water harvesting from various sources such as rooftops, parking lots, and even roads. By harnessing the power of rainwater, preppers can supplement their water supply and reduce their reliance on external sources.

Rooftop rainwater harvesting is a popular method in urban areas. By installing gutters and downspouts, rainwater can be collected and directed into storage containers. This water can then be treated and used for various purposes, including drinking, cooking, and hygiene.

Parking lots and roads can also be utilized for rainwater harvesting. By implementing permeable pavement and underground storage systems, rainwater can be captured and stored for later use. This method not only helps in water conservation but also reduces the strain on the municipal water system during times of crisis.

It is important to note that before implementing any rainwater harvesting system, it is essential to check local regulations and obtain any necessary permits. Some areas may have restrictions or guidelines regarding the collection and use of rainwater.

By exploring and utilizing these urban water sources, preppers can enhance their water resilience and ensure a more sustainable water supply during emergencies. Remember, being prepared is not just about stockpiling water but also understanding the available resources and being adaptable in challenging situations.

5.3 Alternative Water Sources

In times of crisis, when traditional water sources may become compromised or inaccessible, it is crucial for preppers to explore alternative water sources. These alternative sources can provide a lifeline when the usual water supply systems fail. In this chapter, we will delve into various alternative water sources, including ponds, wells, and cisterns, as well as unconventional sources like dew and plants. By understanding and harnessing these sources, you can ensure a more sustainable water supply for you and your loved ones during emergencies.

5.3.1 Ponds: Nature's Reservoirs

Ponds can be a valuable source of water during a crisis. These small bodies of water can be found in various locations, such as parks, forests, and even private properties. Ponds are often fed by natural sources like rainwater, springs, or streams, making them relatively reliable in terms of water availability. However, it is essential to assess the quality of the water before consumption.

To utilize a pond as a water source, you must first ensure that the water is safe to drink. Contaminants such as bacteria, parasites, and chemicals can pose a significant risk to your health. Therefore, it is crucial to treat the water properly before consumption. Boiling, chemical treatment, or filtration systems can be used to purify pond water and make it safe for drinking.

5.3.2 Wells: Tapping into the Earth's Hidden Treasure

Wells have been a reliable source of water for centuries, and they continue to be a valuable asset during emergencies. If you have access to a well, you have a self-sustaining water source that can provide for your needs without relying on external systems. However, it is essential to understand the different types of wells and their maintenance requirements.

There are two main types of wells: shallow wells and deep wells. Shallow wells are typically less than 30 feet deep and are more susceptible to contamination from surface water. Deep wells, on the other hand, can reach depths of hundreds of feet and are less likely to be affected by surface contaminants. Regular testing and maintenance of wells are crucial to ensure the water remains safe for consumption.

In addition to maintaining the well, it is essential to have a backup power source, such as a generator, to operate the pump during power outages. Without electricity, accessing the water from the well becomes challenging. By having a backup power source, you can ensure a continuous supply of water even during extended periods of crisis.

5.3.3 Cisterns: Storing Water for the Long Haul

Cisterns are large containers designed to store rainwater or other water sources for future use. They can be above ground or underground and come in various sizes to accommodate different water storage needs. Cisterns are an excellent option for preppers who want to have a long-term water supply without relying solely on natural sources.

To effectively use a cistern as an alternative water source, it is crucial to consider the quality of the water being stored. Rainwater, for example, is generally safe for non-potable uses but may require treatment before drinking or cooking. Regular maintenance of the cistern, including cleaning and disinfection, is essential to prevent the growth of bacteria or algae.

When installing a cistern, it is important to consider factors such as location, accessibility, and protection from contamination. Placing the cistern in a shaded area can help prevent the growth of algae, while ensuring it is easily accessible for maintenance and water extraction is vital. Additionally, implementing a filtration system or other water treatment methods can further enhance the quality of the stored water.

5.3.4 Unconventional Sources: Dew, Plants, and More

In dire situations where traditional water sources are scarce or unavailable, it is essential to explore unconventional sources of water. Nature provides us with ingenious ways to extract water, even in the most challenging environments. Dew, for example, can be collected by placing a clean cloth or plastic sheet in an open area overnight. The collected dew can then be wrung out and used for drinking or other essential purposes.

Plants can also serve as a source of water in emergencies. Certain plants, such as cacti and succulents, store water in their tissues, making them potential sources of hydration. However, it is crucial to research and identify the plants that are safe to consume and understand the proper methods of extracting water from them.

While these unconventional sources may not provide a significant amount of water, they can supplement your existing water supply and help you survive until more substantial sources become available.

By exploring alternative water sources such as ponds, wells, cisterns, and unconventional sources like dew and plants, you can expand your options for obtaining water during a crisis. Remember to always prioritize water safety and employ appropriate treatment methods to ensure the water you consume is free from contaminants. With a diverse range of water sources at your disposal, you can enhance your preparedness and increase your chances of survival in challenging situations.

5.4 Water from Unconventional Sources

Water is a precious resource, especially in times of crisis. As a prepper, it's crucial to be prepared for any water-related emergency. While we have already explored various conventional sources of water, such as rivers, lakes, and wells, there are also unconventional sources that can provide us with this life-sustaining liquid. In this chapter, we will dive into the world of unconventional water sources, including dew, plants, and more. So, let's explore these unique sources and learn how to extract water from them when traditional sources are scarce.

5.4.1 Dew: Nature's Gift

When you wake up early in the morning and step outside, you may notice tiny droplets of water on the grass or leaves. This is dew, and it can be a valuable source of water in certain situations. Dew forms when the temperature of an object, such as a plant or the ground, drops below the dew point temperature. As the air cools, it loses its ability to hold moisture, causing water vapor to condense into liquid form on surfaces.

To collect dew, you can use a variety of methods. One simple technique is to place a clean, non-absorbent material, such as a plastic sheet or tarp, on the ground overnight. In the morning, the dew will have condensed on the surface of the material. Carefully lift the material and collect the water droplets in a container. It's important to note that dew collection works best in areas with high humidity and cooler temperatures.

5.4.2 Plants: Nature's Reservoirs

Plants are not only essential for our survival but can also serve as a source of water in times of need. Certain plants, such as cacti and succulents, have adapted to arid environments by storing water in their tissues. By tapping into these natural reservoirs, we can extract water for our own use.

To extract water from plants, you can follow these steps:

1. Identify plants with water-storing capabilities, such as cacti or certain types of agave.

2. Use a sharp knife or blade to carefully cut into the plant's flesh.

3. Collect the liquid that oozes out of the plant and store it in a clean container.

4. Filter the collected liquid to remove any impurities before consumption.

It's important to remember that not all plants are safe to consume, and some may even be toxic. Therefore, it's crucial to research and identify the specific plants in your area that can provide safe drinking water.

While it is important to research and identify specific plants in your area, here is a list of commonly known plants that are generally safe to consume for extracting water:

- Cacti (such as prickly pear cactus)
- Agave
- Bamboo
- Banana plant
- Coconut palm
- Pineapple
- Aloe vera
- Saguaro cactus
- Yucca
- Bamboo palm

Remember, it is crucial to have proper knowledge and identification of plants before consuming any part of them.

5.4.3 Transpiration: Nature's Water Cycle

Transpiration is a natural process by which plants release water vapor through their leaves. This process can be harnessed to extract water in certain situations. By creating a simple transpiration bag, you can collect water from plants even in arid environments.

To create a transpiration bag, follow these steps:

1. Select a leafy branch from a non-toxic plant.
2. Enclose the branch in a clear plastic bag, ensuring that the bag is tightly sealed around the stem.
3. Secure the bag in place, making sure it is exposed to sunlight.
4. Over time, water vapor will accumulate inside the bag as the plant transpires.
5. Carefully remove the bag and collect the condensed water from the inside.

This method may not yield large quantities of water, but it can provide a valuable source of hydration in emergency situations.

5.4.4 Other Unconventional Sources

In addition to dew and plants, there are other unconventional sources of water that can be explored in times of need. These include:

- **Air conditioning units**: Condensation from air conditioning units can be collected and used for non-potable purposes, such as cleaning or flushing toilets.
- **Household items**: Everyday items, such as towels or clothing, can be used to absorb

moisture from the environment. By wringing out these items, you can collect the water for various uses.

- **Underground water sources**: In certain areas, water can be found underground by digging or using specialized equipment. However, accessing underground water sources should be done with caution and proper knowledge.

Remember, these unconventional sources should be considered as a last resort when traditional sources are unavailable or contaminated. It's essential to prioritize safety and ensure proper treatment and filtration of water before consumption.

In the next chapter, we will explore the critical topic of water storage and containers. Understanding how to store water safely and efficiently is vital for long-term survival. So, let's dive into the world of water storage and learn how to ensure a reliable water supply in times of crisis.

Water Storage and Containers

6.1 Choosing the Right Water Storage Containers

Water storage is a critical aspect of emergency preparedness. When a crisis strikes and water becomes scarce, having a reliable and safe water supply can mean the difference between survival and desperation. In this chapter, we will explore the importance of choosing the right water storage containers and provide practical guidance on how to do so effectively.

Understanding the Importance of Proper Water Storage

Before we delve into the specifics of choosing water storage containers, it is crucial to understand why proper water storage is essential. In times of crisis, access to clean and safe drinking water may be limited or completely cut off. Having an adequate supply of stored water ensures that you and your loved ones can stay hydrated and maintain good health.

When selecting water storage containers, there are several factors to consider. These include the material of the container, its capacity, durability, and ease of use. Let's explore each of these aspects in detail.

Material of the Container

The material of the water storage container plays a significant role in ensuring the safety and longevity of your stored water. Here are some common materials used for water storage containers:

- **Plastic**: Food-grade plastic containers, such as high-density polyethylene (HDPE) or polypropylene (PP), are popular choices for water storage. These containers are lightweight, durable, and resistant to chemicals and UV radiation. Look for containers labeled as "food-grade" or "safe for drinking water" to ensure they meet the necessary standards.
- **Glass**: Glass containers are an excellent option for long-term water storage. They are non-permeable, meaning they won't leach any harmful substances into the water. However, glass containers can be heavy and prone to breakage, so they may not be suitable for all situations.
- **Stainless Steel**: Stainless steel containers are highly durable and resistant to corrosion. They are a good choice for long-term storage

and can withstand extreme temperatures. However, stainless steel containers can be more expensive than other options.
- **Ceramic**: Ceramic containers are aesthetically pleasing and can keep water cool. However, they are fragile and may not be suitable for emergency situations where durability is crucial.

Capacity of the Container

Determining the appropriate capacity for your water storage containers depends on several factors, including the number of people in your household and the duration of the emergency. As a general guideline, it is recommended to store at least one gallon (3.8 liters) of water per person per day for drinking and sanitation purposes. This amount should be sufficient for a minimum of three days, but it is advisable to store more if possible.

Consider the space available for storage when deciding on the container capacity. It may be more practical to have several smaller containers rather than one large container, as this allows for easier transportation and distribution of water if needed.

Durability and Ease of Use

In an emergency situation, you need water storage containers that are durable and easy to handle. Look for containers that are designed to withstand rough handling and can be easily sealed to prevent contamination. Some containers come with built-in handles or spigots, making it convenient to access and dispense water.

Consider the space available for storage when deciding on the container capacity. It may be more practical to have several smaller containers rather than one large container, as this allows for easier transportation and distribution of water if needed..

Summary

Choosing the right water storage containers is crucial for ensuring a reliable and safe water supply during emergencies. Consider the material, capacity, durability, and ease of use when selecting containers. Plastic containers, such as HDPE or PP, are popular choices due to their durability and resistance to chemicals. Glass containers are suitable for long-term storage but may be heavy and prone to breakage. Stainless steel containers offer durability and resistance to corrosion. Determine the appropriate capacity based on the number of people and the duration of the emergency. Finally, prioritize durability and ease of use to ensure your water storage containers can withstand the challenges of a crisis situation.

6.2 Water Storage Guidelines

Water storage is a crucial aspect of emergency preparedness. In times of crisis, access to clean and safe water may become limited or even nonexistent. As a prepper, it is essential to have a sufficient supply of water stored to ensure the well-being and survival of yourself and your loved ones. But how much water do you actually need? In this section, we will explore water storage guidelines to help you determine the right amount of water to store for various emergency situations.

6.2.1 Determining Your Water Needs

Before diving into the specifics of water storage, it is important to understand how much water you and your family require on a daily basis. The general rule of thumb is to store at least one gallon of water per person per day. This includes both drinking water and water for hygiene and sanitation purposes. However, certain factors may influence your water needs, such as climate, age, physical activity level, and the presence of medical conditions.

To calculate your water needs more accurately, consider the following guidelines:

- **Drinking Water**: Each person should have a minimum of one gallon of water per day for drinking purposes. This will ensure hydration and help maintain overall health.
- **Food Preparation**: If you plan to cook meals during an emergency, you will need additional water for food preparation. Estimate around one gallon per day for cooking and food hygiene.
- **Hygiene and Sanitation**: Adequate water for personal hygiene and sanitation is crucial to prevent the spread of diseases. Allocate at least one gallon per day for activities such as handwashing, bathing, and dishwashing.
- **Medical Needs**: If you or any family members have specific medical conditions that require additional water, consult with a healthcare professional to determine the appropriate amount.

Remember, these guidelines are just a starting point. It is always better to err on the side of caution and store more water than you think you will need. In an emergency, you may not have access to alternative water sources for an extended period.

6.2.2 Storing Water Safely

Once you have determined how much water you need to store, it is crucial to ensure that the water remains safe and potable for an extended period. Here are some guidelines for safe water storage:

- **Choose the Right Containers**: Select food-grade containers specifically designed for water storage. Avoid using containers that previously held chemicals or non-food items, as they may contaminate the water. Plastic containers, such as BPA-free water jugs or food-grade buckets with tight-fitting lids, are commonly used for water storage.
- **Clean and Disinfect Containers**: Before filling the containers with water, thoroughly clean them with mild soap and water. Rinse them well to remove any residue. To disinfect the containers, use a solution of one teaspoon of unscented household bleach per gallon of water. Allow the containers to air dry before filling them.
- **Fill Containers with Water**: Fill the containers with tap water, leaving about an inch of headspace to allow for expansion if the water freezes. If tap water is not available or is of questionable quality, consider using

commercially bottled water or properly treated and filtered water from alternative sources.

- **Label and Date Containers**: Clearly label each container with the date of filling and store them in a cool, dark place away from direct sunlight. This will help you keep track of the water's freshness and rotation.
- **Rotate Water**: Regularly check the stored water and rotate it every six months to ensure freshness. Use the older water for non-potable purposes, such as cleaning or gardening, and refill the containers with fresh water.

By following these guidelines, you can ensure that your stored water remains safe and ready for use during emergencies. Remember to regularly review and update your water storage plan to account for any changes in your family's needs or circumstances.

6.2.3 Additional Considerations

While storing water in containers is the most common method, there are alternative solutions to consider for emergency water storage:

- **Water Bladders**: Flexible water bladders are an excellent option for those with limited

storage space. These bladders can be placed in bathtubs or other suitable areas and filled with water as needed.
- **Rainwater Harvesting**: If you have a rainwater harvesting system in place, consider storing excess rainwater in large storage tanks or barrels. Ensure that the collected rainwater is properly filtered and treated before consumption.
- **Swimming Pools**: In certain emergency situations, swimming pools can serve as a temporary water source. However, it is important to note that pool water may contain chemicals and should be properly treated before use.

Remember, water is a precious resource, and having a sufficient supply during emergencies is vital. By following these water storage guidelines and regularly maintaining your water storage plan, you can ensure the well-being and survival of yourself and your loved ones in times of crisis. Stay prepared, stay safe!

6.3 Water Rotation

Water rotation is a crucial aspect of maintaining a fresh and safe water supply in times of crisis. As a prepper, you understand the importance of having a sufficient amount of water stored for emergencies. However, it is equally important to ensure that the water you store remains fresh and safe to consume. In this section, we will explore the concept of water rotation and provide you with practical tips on how to effectively manage your water supply.

Why is Water Rotation Important?

Imagine this scenario: you have diligently stored water for an emergency, but when the time comes to use it, you discover that the water has become stagnant, foul-smelling, and potentially unsafe to drink. This is where water rotation becomes crucial. By regularly rotating your water supply, you can ensure that the water remains fresh, free from contaminants, and safe for consumption.

Water rotation serves two main purposes. Firstly, it helps to prevent the growth of bacteria, algae, and other microorganisms that can thrive in stagnant water. Secondly, it allows you to regularly inspect and assess the quality of your stored water, ensuring that it is still suitable for use.

How Often Should You Rotate Your Water?

The frequency of water rotation depends on various factors, including the type of containers used, the storage conditions, and the quality of the water source. As a general guideline, it is recommended to rotate your water supply every six months. However, it is essential to regularly inspect your water containers and assess the quality of the water to determine if rotation is necessary sooner.

Inspecting and Testing Your Water

Regularly inspecting and testing your stored water is crucial to ensure its safety and quality. Here are some steps you can take to assess the condition of your water:

- **Visual Inspection**: Check for any signs of discoloration, sediment, or floating particles in the water. If you notice any of these, it may indicate contamination or degradation of the water quality.
- **Odor Test**: Take a whiff of the water. If it has a foul or unusual smell, it is a clear indication that the water has become contaminated and should not be consumed.
- **Taste Test**: While it is not recommended to taste water directly from your storage containers, you can pour a small amount into a clean glass and taste it. If the water tastes off or has an unpleasant flavor, it is best to discard it.

- **Water Testing Kits**: Consider investing in water testing kits that can provide more accurate results. These kits can detect the presence of harmful bacteria, viruses, and other contaminants in your water supply.

Rotating Your Water Supply

When it comes to rotating your water supply, there are a few methods you can employ:

- **Consumption Rotation**: One of the simplest ways to rotate your water is by incorporating it into your daily life. Use the stored water for drinking, cooking, and other household needs, ensuring that you regularly replenish your supply with fresh water.
- **Donation or Sharing**: If you have a surplus of stored water that is nearing its expiration date, consider donating it to local charities, emergency response organizations, or neighbors in need. This not only helps others but also ensures that your water supply remains fresh.
- **Gardening and Irrigation**: Utilize your stored water for gardening and irrigation purposes. Water your plants, vegetables, or fruit trees with the stored water, allowing you to rotate and replenish your supply regularly.

- **Watering Livestock or Pets**: If you have animals or pets that require water, use your stored water to meet their needs. This not only helps with water rotation but also ensures that your animals have access to clean water during emergencies.

Maintaining Water Storage Containers

Proper maintenance of your water storage containers is essential to prevent contamination and ensure the longevity of your water supply. Here are some tips for maintaining your containers:

- **Cleanliness**: Regularly clean your water storage containers with mild soap and water. Rinse them thoroughly to remove any residue or cleaning agents before refilling them with fresh water.
- **Sanitization**: Periodically sanitize your containers to eliminate any potential bacteria or contaminants. You can use a diluted bleach solution or other sanitizing agents recommended for water storage containers.
- **Sealing and Protection**: Ensure that your containers are tightly sealed to prevent the entry of insects, dust, or other contaminants. Store them in a cool, dry place away from

direct sunlight to maintain the quality of the water.
- **Labeling and Dating**: Label your water storage containers with the date of filling and rotation to keep track of the freshness of the water. This will help you prioritize the use of older water and ensure that it is consumed or rotated before newer supplies.

By following these practices, you can maintain a fresh and safe water supply for your emergency needs. Remember, water rotation is not just about having water stored; it is about ensuring that the water remains viable and safe for consumption. Stay vigilant, regularly inspect your water supply, and rotate it accordingly to be well-prepared for any water-related crisis that may come your way.

6.4 Emergency Water Storage Solutions

In times of crisis, having a reliable and sufficient water supply is crucial for survival. As a prepper, you understand the importance of being prepared for any emergency, including water-related ones. While traditional water storage containers are essential, there are also alternative solutions that can provide you with additional options for storing water in an emergency. In this section, we will explore some of these emergency water storage solutions that go beyond the traditional containers.

6.4.1 Water Bladders and Tanks

Water bladders and tanks are flexible containers that can hold a large volume of water. They are designed to be durable, portable, and easy to store. These containers are typically made from materials such as PVC or polyethylene, which are resistant to punctures and UV rays. Water bladders and tanks come in various sizes, ranging from small portable options to large tanks that can hold thousands of gallons of water.

One advantage of using water bladders and tanks is their versatility. They can be easily transported and set up in different locations, making them ideal for both temporary and long-term water storage. These containers can be placed in basements, garages, or even outdoors, as long as they are protected from direct sunlight and extreme temperatures.

When using water bladders and tanks, it is important to ensure that they are properly cleaned and sanitized before filling them with water. Regular maintenance and inspection are also necessary to prevent leaks or damage. Additionally, it is crucial to have a reliable method of accessing the water stored in these containers, such as a gravity-fed system or a pump.

6.4.2 Rain Barrels and Cisterns

Rain barrels and cisterns are excellent options for emergency water storage, especially in areas where rainfall is abundant. These systems collect and store rainwater from rooftops, which can then be used for various purposes, including drinking, cooking, and gardening.

Rain barrels are typically smaller in size and can be easily installed under downspouts to capture rainwater. They often come with a spigot for easy access to the stored water. Cisterns, on the other hand, are larger storage containers that can hold a significant amount of rainwater. They are usually installed underground or above ground and require a more complex setup.

When using rain barrels and cisterns, it is important to ensure that the collected rainwater is properly filtered and treated before consumption. This can be done through various methods, such as using a filtration system or adding disinfectants like chlorine. Regular maintenance, including cleaning and inspecting the containers, is also necessary to prevent contamination and ensure the quality of the stored water.

6.4.3 Underground Water Storage

In some situations, it may be necessary to store water underground to protect it from external factors such as temperature fluctuations or potential contamination. Underground water storage options include buried tanks, wells, and natural underground reservoirs.

Buried tanks are specifically designed for underground water storage. They are made from materials that can withstand the pressure of the surrounding soil and are equipped with proper ventilation and access points. These tanks can be installed in basements or specially constructed underground chambers.

Wells are another option for underground water storage. They tap into natural underground water sources and provide a continuous supply of water. However, it is important to ensure that the well is properly constructed, maintained, and protected from contamination.

Natural underground reservoirs, such as caves or natural rock formations, can also be used for water storage. These reservoirs often require additional measures to ensure the water's safety and accessibility, such as proper sealing and filtration systems.

When considering underground water storage, it is crucial to consult with professionals who have expertise in well drilling, tank installation, or underground construction. They can provide guidance on the best practices and ensure that the storage system is safe and reliable.

6.4.4 Portable Water Filtration Systems

In addition to traditional water storage containers, portable water filtration systems can be a valuable asset in emergency situations. These systems are designed to remove contaminants and purify water from various sources, including rivers, lakes, and even questionable tap water.

Portable water filtration systems come in different forms, such as straw filters, pump filters, and gravity filters. They are compact, lightweight, and easy to carry, making them ideal for on-the-go situations or when traditional water sources are not available.

When using portable water filtration systems, it is important to understand their limitations and capabilities. Different filters may have varying levels of effectiveness in removing specific contaminants, so it is crucial to choose a system that suits your needs and the water sources you are likely to encounter.

Regular maintenance and cleaning of portable water filtration systems are essential to ensure their optimal performance. It is also important to have spare parts and backup filters on hand in case of emergencies or extended use.

By incorporating these emergency water storage solutions into your preparedness plan, you can enhance your ability to secure and maintain a reliable water supply during times of crisis. Remember to regularly inspect, clean, and maintain these alternative storage options to ensure their effectiveness and longevity. With a well-thought-out water storage strategy, you can face any water-related emergency with confidence and resilience.

Water for Hygiene and Sanitation

7.1 Maintaining Personal Hygiene in Water-Scarce Situations

In a water-scarce situation, personal hygiene becomes a critical aspect of survival. While finding and treating water for drinking and cooking is essential, it is equally important to maintain cleanliness to prevent the spread of diseases and infections. In this chapter, we will explore various techniques and strategies to help you maintain personal hygiene when water is limited. From alternative cleaning methods to conserving water for hygiene purposes, we will equip you with the knowledge and skills necessary to stay clean and healthy in crisis situations.

7.1.1 The Importance of Personal Hygiene

Personal hygiene plays a vital role in preventing the spread of diseases and maintaining overall well-being. In a water-scarce situation, where access to clean water is limited, practicing good hygiene becomes even more crucial. Poor hygiene can lead to the proliferation of bacteria, viruses, and parasites, increasing the risk of infections and illnesses. By prioritizing personal hygiene, you can significantly reduce the chances of falling ill and ensure your overall health and well-being.

7.1.2 Waterless Hygiene Alternatives

When water is scarce, it is essential to explore alternative methods of maintaining personal hygiene. Waterless hygiene alternatives can help conserve water while still allowing you to stay clean and fresh. One such alternative is the use of alcohol-based hand sanitizers. These sanitizers effectively kill germs and bacteria on your hands without the need for water. Additionally, wet wipes or pre-moistened towelettes can be used for cleaning your body when water is not readily available. These alternatives can be a lifesaver in situations where water is scarce or contaminated.

7.1.3 Creative Hygiene Solutions

In water-scarce situations, it is crucial to think creatively and find innovative solutions to maintain personal hygiene. One such solution is the use of dry shampoo. Dry shampoo is a powder or spray that absorbs excess oil and dirt from your hair, allowing you to keep it clean without the need for water. Another creative solution is the use of baking soda as a natural deodorant. Baking soda helps neutralize odors and can be applied directly to your underarms to keep them fresh and odor-free. These creative hygiene solutions can be a game-changer when water is scarce.

7.1.4 Prioritizing Hygiene Practices

In a crisis situation, it is important to prioritize hygiene practices to ensure the health and well-being of yourself and your loved ones. Start by establishing a routine for personal hygiene and stick to it. This routine should include regular handwashing, even if water is limited. When water is scarce, use it sparingly and prioritize its use for drinking and cooking. Explore alternative cleaning methods and make use of waterless hygiene alternatives whenever possible. By prioritizing hygiene practices, you can minimize the risk of infections and maintain good health in water-scarce situations.

7.2 Sanitation Practices

In a crisis situation, ensuring proper sanitation practices becomes crucial for preventing the spread of waterborne diseases. When water sources are scarce or contaminated, maintaining good hygiene and sanitation can be challenging. However, with the right knowledge and preparation, you can minimize the risk of illness and keep yourself and your loved ones safe.

7.2.1 The Importance of Sanitation

In times of crisis, access to clean water for hygiene and sanitation purposes may be limited. Without proper sanitation practices, the risk of waterborne diseases such as cholera, dysentery, and typhoid fever increases significantly. These diseases can spread rapidly and have severe consequences for individuals and communities.

Sanitation practices are essential for preventing the contamination of water sources and reducing the transmission of diseases. By implementing proper sanitation measures, you can protect yourself and others from illness, maintain a clean living environment, and promote overall well-being.

7.2.2 Hygiene and Sanitation in Water-Scarce Situations

When water is scarce, maintaining personal hygiene can be challenging. However, there are several strategies you can employ to make the most of limited water resources. These include:

- **Waterless hygiene alternatives**: In situations where water is scarce, it is essential to conserve water for drinking and cooking. Waterless hygiene alternatives, such as using hand sanitizers, dry shampoo, and wet wipes, can help you maintain personal hygiene without relying on large amounts of water.

- **Conserving water**: Implementing water conservation techniques, such as taking shorter showers, turning off the tap while brushing your teeth, and using water-saving devices, can help stretch your water supply further. By reducing water waste, you can ensure that you have enough water for essential hygiene practices.

- **Reusing water**: Greywater recycling, which involves collecting and treating water from sources such as showers, sinks, and washing machines, can be used for non-potable purposes such as flushing toilets or watering plants. By reusing water, you can minimize water wastage and maximize its utility.

7.2.3 Preventing Waterborne Diseases

Preventing waterborne diseases requires a combination of proper sanitation practices and water treatment techniques. Here are some key steps to follow:

- **Handwashing**: Regular and thorough handwashing with soap and clean water is one of the most effective ways to prevent the spread of waterborne diseases. Make sure to wash your hands before handling food, after using the toilet, and whenever they are visibly dirty.

- **Proper waste disposal**: Dispose of waste properly to prevent contamination of water sources. Use designated waste disposal areas or facilities and avoid throwing waste into rivers, lakes, or other water bodies.

- **Disinfection**: If you are unsure about the safety of the water you have access to, it is crucial to disinfect it before use. Boiling, chemical treatment, filtration, or solar disinfection methods discussed in earlier chapters can help eliminate harmful pathogens and make the water safe for consumption and sanitation purposes.

- **Maintaining clean living spaces**: Keep your living spaces clean and free from potential sources of contamination. Regularly clean surfaces, utensils, and food preparation areas with clean water and soap or disinfectant.

7.2.4 Emergency Toilet Options

In a crisis situation, proper sanitation facilities may not be readily available. It is essential to have a plan in place for dealing with sanitation challenges. Here are some emergency toilet options to consider:

- **Portable toilets**: Portable toilets, such as camping toilets or portable commodes, can provide a temporary solution for managing human waste. These toilets are designed to be easily transportable and can be set up in a designated area away from water sources.

- **Bucket toilets**: A simple bucket toilet can be created by placing a plastic bag or a bucket with a toilet seat on top. After use, the waste can be sealed in the bag and disposed of properly.

- **Composting toilets**: Composting toilets are a more sustainable long-term solution for managing human waste. These toilets use natural processes to break down waste into compost that can be safely used as fertilizer.

- **Digging latrines**: In situations where you have access to land, digging a latrine can provide a more permanent solution for waste disposal. Ensure that the latrine is located at a safe distance from water sources and follow proper guidelines for construction and maintenance.

Remember, proper sanitation practices are essential for maintaining health and preventing the spread of waterborne diseases. By implementing these practices and having a plan in place for emergency sanitation, you can ensure the well-being of yourself and your loved ones even in challenging circumstances.

7.3 Waterless Hygiene Alternatives

In a crisis situation where water is scarce, maintaining personal hygiene becomes a challenge. However, it is crucial to prioritize hygiene to prevent the spread of diseases and maintain overall well-being. In this chapter, we will explore waterless hygiene alternatives that can help you stay clean and healthy even when water is limited. These alternatives are practical, effective, and can make a significant difference in your daily life during an emergency.

7.3.1 Dry Shampoo: Keeping Your Hair Fresh

When water is scarce, washing your hair may seem like a luxury you can't afford. However, with the help of dry shampoo, you can keep your hair clean and fresh without using water. Dry shampoo is a powder or spray that absorbs excess oil and dirt from your hair, leaving it looking and feeling clean. To use dry shampoo, simply apply it to your roots, massage it in, and brush out any residue. Not only does dry shampoo save water, but it also saves time and energy, making it an excellent option for maintaining personal hygiene in water-scarce situations.

7.3.2 Wet Wipes: A Quick Refresh

Wet wipes are a versatile and convenient waterless hygiene alternative. They are pre-moistened disposable wipes that can be used for cleaning your body, hands, and face. Wet wipes are especially useful when water is limited or not readily available. They are easy to carry in your emergency kit or bug-out bag and can provide a quick refresh when you need it the most. Whether you're camping, traveling, or facing a water-related emergency, wet wipes can help you stay clean and hygienic.

7.3.3 Hand Sanitizer: Killing Germs on the Go

Hand sanitizer is an essential item in any emergency preparedness kit. When water is scarce, hand sanitizer can be a lifesaver for maintaining hand hygiene. It is a gel or liquid that contains alcohol, which kills germs and bacteria on your hands. To use hand sanitizer, simply apply a small amount to your hands and rub them together until dry. Hand sanitizer is particularly useful in situations where soap and water are not available, such as during power outages or when you're on the move. Remember to choose a hand sanitizer with at least 60% alcohol content for maximum effectiveness.

7.3.4 Dry Bathing: Staying Fresh without Water

Dry bathing is a technique that allows you to clean your body without using water. It involves using absorbent materials, such as towels or wipes, along with cleansing products to remove dirt and odor from your skin. To dry bathe, start by applying a small amount of body wash or soap to a towel or wipe. Then, gently rub the towel or wipe over your body, focusing on areas that tend to accumulate sweat and odor. Dry bathing not only helps you stay clean but also provides a refreshing feeling, especially in hot and humid environments.

7.3.5 Oral Hygiene: Freshening Your Breath

Maintaining oral hygiene is essential for overall health, even in water-scarce situations. When water is limited, you can still keep your mouth clean and fresh with waterless oral hygiene alternatives. One option is to use mouthwash or mouth rinse that does not require water. These products contain antibacterial agents that help kill germs and freshen your breath. Another option is to use oral hygiene wipes or dental foam, which can be applied directly to your teeth and gums for cleaning. Remember to brush your teeth regularly, even if you don't have access to water, as it plays a vital role in preventing dental issues.

7.3.6 Powdered Cleansers: Multipurpose Cleaning

Powdered cleansers are versatile waterless hygiene alternatives that can be used for various cleaning purposes. These cleansers come in the form of powders or granules and can be mixed with a small amount of water or used dry. They are effective for cleaning surfaces, dishes, and even your body. Powdered cleansers often contain mild abrasives and cleaning agents that help remove dirt, grease, and stains. They are an excellent option for maintaining cleanliness and hygiene when water is limited.

7.3.7 Essential Oils: Natural Freshness

Essential oils can be a valuable addition to your waterless hygiene routine. They not only provide a pleasant fragrance but also possess antibacterial and antifungal properties. You can use essential oils to freshen your body, clothes, and living space. Simply add a few drops of your favorite essential oil to a cloth or tissue and gently rub it on your skin or place it in your surroundings. Lavender, tea tree, and eucalyptus essential oils are particularly known for their cleansing and refreshing properties.

7.3.8 Hygiene Kits: Ready-to-Use Solutions

To make waterless hygiene more convenient, consider assembling hygiene kits that contain all the necessary items for staying clean and fresh. These kits can include items such as dry shampoo, wet wipes, hand sanitizer, oral hygiene products, powdered cleansers, and essential oils. By having a hygiene kit readily available, you can ensure that you have everything you need to maintain personal hygiene during a water-related emergency.

Remember, while waterless hygiene alternatives can be effective in conserving water, it is still essential to prioritize water for drinking and cooking. Use these alternatives wisely and conserve water whenever possible. By incorporating waterless hygiene practices into your emergency preparedness plan, you can stay clean, healthy, and resilient in the face of water scarcity.

7.4 Emergency Toilet Options

When it comes to preparing for a crisis, water is undoubtedly one of the most critical resources to consider. As a prepper, you understand the importance of having a reliable supply of clean water for drinking, cooking, and hygiene. But what about sanitation? In an emergency situation, proper sanitation practices can quickly become a matter of life and death.

Imagine this: you find yourself in the midst of a disaster, where the water supply is contaminated or completely cut off. The toilets in your home are no longer functioning, and you're left with no choice but to find alternative solutions for your bathroom needs. This is where understanding emergency toilet options becomes crucial.

In this section, we will explore various emergency toilet options that can help you maintain proper sanitation and hygiene during a crisis. We will discuss both short-term solutions for immediate needs and long-term strategies for extended water scarcity situations. So, let's dive in and explore the world of emergency toilets!

7.4.1 Portable Camping Toilets

One of the most practical and versatile options for emergency sanitation is a portable camping toilet. These compact and lightweight toilets are designed to be easily transported and set up in any location. They typically consist of a seat, a waste collection container, and a mechanism for odor control.

Portable camping toilets are an excellent choice for short-term emergencies or situations where you need a temporary solution. They are easy to use and maintain, and many models come with disposable waste bags that can be sealed and disposed of safely. These toilets are also ideal for outdoor use, such as camping or during power outages.

7.4.2 DIY Bucket Toilets

In a long-term crisis scenario, where water scarcity is a significant concern, a DIY bucket toilet can be a practical and cost-effective solution. These toilets are simple to assemble using basic materials such as a sturdy bucket, a toilet seat, and a sealing lid.

To create a DIY bucket toilet, you can line the bucket with a heavy-duty garbage bag or use a compostable bag for eco-friendly disposal. Adding a scoop of sawdust or cat litter after each use helps control odors and aids in the decomposition process.

When the bag is full, it can be securely tied and disposed of in an appropriate manner.

7.4.3 Composting Toilets

For those seeking a more sustainable and environmentally friendly option, composting toilets offer an innovative solution. Composting toilets use natural processes to break down human waste into compost, which can then be safely used as fertilizer for non-edible plants.

These toilets typically consist of a separate chamber for solid waste and a urine-diverting system. The solid waste chamber is equipped with a ventilation system that promotes aerobic decomposition, while the urine is collected separately for easier management. Composting toilets require regular maintenance and monitoring to ensure proper functioning and odor control.

7.4.4 Chemical Toilets

Chemical toilets are another option for emergency sanitation, commonly used in portable restrooms or during outdoor events. These toilets use chemicals to break down waste and control odors. They are self-contained units that include a waste collection tank, a flushing mechanism, and a chemical solution to aid in decomposition.

Chemical toilets are relatively easy to use and maintain, making them suitable for short-term emergencies. However, it's important to note that the chemicals used in these toilets may have environmental implications, and proper disposal methods should be followed.

7.4.5 Improvised Solutions

In dire situations where no other options are available, improvisation may be necessary. This could involve using a designated area outdoors for waste disposal, following proper hygiene practices to minimize contamination risks. It's crucial to maintain strict sanitation standards and ensure waste is disposed of safely to prevent the spread of diseases.

Remember, while improvised solutions may be necessary in extreme circumstances, they should only be considered as a last resort. It's essential to prioritize hygiene and sanitation to protect yourself and others from potential health hazards.

In conclusion, when preparing for emergencies, it's vital to consider all aspects of water management, including sanitation. Understanding and planning for emergency toilet options can help ensure the health and well-being of yourself and your loved ones during a crisis. Whether it's portable camping toilets,

DIY bucket toilets, composting toilets, chemical toilets, or improvised solutions, having a plan in place will give you peace of mind and help you navigate through challenging times with dignity and safety.

Water in Extreme Environments

8.1 Surviving in Arid and Desert Environments

In the scorching heat of arid and desert environments, water becomes an even more precious resource. The lack of rainfall and the harsh conditions make finding and conserving water a challenging task. As a prepper, it is crucial to understand the unique challenges of these environments and be prepared to handle water-related emergencies. In this chapter, we will explore strategies and techniques to survive and thrive in arid and desert environments.

8.1.1 Understanding the Challenges

Arid and desert environments are characterized by extremely low precipitation and high evaporation rates. The scarcity of water sources and the intense heat make it difficult to find and retain water. In these environments, dehydration can occur rapidly, leading to serious health risks and even death. It is essential to understand the challenges associated with these environments to effectively prepare for water-related emergencies.

8.1.2 Finding Water Sources

In arid and desert environments, finding water sources can be a daunting task. However, with the right knowledge and skills, it is possible to locate hidden water sources. We will discuss various techniques such as searching for signs of vegetation, animal activity, and geological formations that indicate the presence of water.

8.1.3 Collecting and Storing Water

Once you have located a water source, it is crucial to collect and store water efficiently. We will explore different methods of water collection, such as using natural depressions, constructing solar stills, and utilizing condensation techniques. Practical examples and step-by-step instructions will guide you through the process of collecting and storing water in arid and desert environments.

8.1.4 Water Treatment Techniques

Water found in arid and desert environments may be contaminated and unsafe for consumption. We will delve into various water treatment techniques, including boiling, chemical treatment, and filtration. Detailed explanations and practical examples will help you understand how to effectively treat water to make it safe for drinking and cooking.

8.1.5 Water Conservation Strategies

Conserving water is of utmost importance in arid and desert environments. We will discuss practical tips and techniques for minimizing water usage, such as using water-efficient appliances, implementing xeriscaping techniques, and practicing water-wise habits.

8.1.6 Planning for Long-Term Water Scarcity

In arid and desert environments, long-term water scarcity is a significant concern. We will explore strategies for coping with extended water shortages, such as rainwater harvesting, greywater recycling, and implementing water-efficient irrigation techniques. Additionally, we will discuss the importance of community water solutions and creating a personal water emergency plan to ensure long-term water security.

8.2 Water Challenges in Cold Climates

In extreme environments, such as cold climates, water becomes even more crucial for survival. The freezing temperatures and icy conditions pose unique challenges when it comes to finding, treating, and storing water. As a prepper, it is essential to understand these challenges and be prepared to overcome them in order to ensure your water security in cold climates.

8.2.1 The Importance of Water in Cold Climates

Water is vital for survival in any environment, but in cold climates, it becomes even more critical. The human body requires water to regulate temperature, maintain bodily functions, and prevent dehydration. In cold weather, the body loses water through respiration, sweating, and increased urine production. Additionally, the dry air in cold climates can cause increased water loss through evaporation from the skin.

Furthermore, access to clean water is essential for preventing hypothermia. In cold climates, the body loses heat more rapidly, and staying hydrated helps regulate body temperature and prevent frostbite and other cold-related injuries.

8.2.2 Snow and Ice as Water Sources

In cold climates, snow and ice can serve as valuable water sources. However, it is important to understand how to collect and treat snow and ice to ensure its safety for consumption.

Collecting Snow

Collecting snow is a simple and effective way to obtain water in cold climates. However, it is crucial to collect clean snow that is free from contaminants. Avoid collecting snow from areas near roads, industrial sites, or heavily polluted areas.

To collect snow, use a clean container or fabric to scoop it up. Melt the snow by placing the container near a heat source or using body heat. It is important to note that snow has a lower water content than ice, so you will need a larger volume of snow to obtain a sufficient amount of water.

Melting Ice

Ice can also be melted to obtain water in cold climates. However, it is essential to ensure that the ice is free from contaminants before melting it. Avoid using ice that has come into contact with pollutants or has an unusual color or odor.

To melt ice, place it in a clean container and allow it to thaw naturally or use a heat source. It is important to monitor the melting process to prevent the ice from refreezing and to collect the liquid water as it melts.

8.2.3 Treating Snow and Ice for Consumption

While snow and ice can be a source of water in cold climates, they may contain impurities and pathogens that can cause illness. It is crucial to treat snow and ice before consuming it to ensure its safety.

Boiling

Boiling is one of the most reliable methods for treating water, even in cold climates. Bring the melted snow or ice to a rolling boil for at least one minute to kill any harmful bacteria, viruses, and parasites. Boiling water will also help melt any remaining ice or snow particles.

Chemical Treatment

Chemical treatment methods, such as using water purification tablets or liquid disinfectants, can also be used to treat water in cold climates. Follow the instructions provided with the chemical treatment product to ensure effective disinfection.

Filtration

Filtration systems can be used to remove impurities and particles from melted snow or ice. Look for water filters specifically designed for cold weather conditions, as some filters may become less effective in freezing temperatures. It is important to regularly clean and maintain the filter to ensure its proper functioning.

8.2.4 Storing Water in Cold Climates

Storing water in cold climates requires special considerations to prevent freezing and ensure a continuous water supply.

Insulation

To prevent water from freezing, insulate your water storage containers by wrapping them in insulating materials such as foam or blankets. This will help maintain the temperature and prevent freezing.

Underground Storage

In extremely cold climates, consider burying your water storage containers underground. The earth's natural insulation will help prevent freezing and keep the water at a more stable temperature.

Heat Sources

If possible, store your water containers near a heat source, such as a wood stove or heater. The heat will help prevent freezing and ensure a readily available water supply.

Regular Monitoring

Regularly check your water storage containers for signs of freezing or damage. If any containers have frozen, allow them to thaw naturally before using the water. It is also important to rotate your water supply regularly to ensure freshness and prevent stagnation.

In cold climates, water is a precious resource that requires careful management and preparation. By understanding the unique challenges and implementing the appropriate techniques for finding, treating, and storing water, you can ensure your water security even in the harshest of conditions. Stay hydrated, stay safe, and be prepared for any water-related emergencies that may arise in cold climates.

8.3 Water in Coastal and Marine Environments

Water is a precious resource that is essential for our survival. As preppers, we understand the importance of being prepared for any emergency situation, and that includes being equipped to handle water-related challenges. In this chapter, we will explore the unique considerations and techniques for obtaining and utilizing water in coastal and marine environments.

8.3.1 Understanding the Challenges

Coastal and marine environments present their own set of challenges when it comes to accessing and utilizing water. The abundance of saltwater and the harsh conditions can make it difficult to find safe and drinkable water. However, with the right knowledge and techniques, it is possible to overcome these challenges and ensure a reliable water supply.

8.3.2 Desalination Techniques

Desalination is the process of removing salt and other impurities from seawater, making it safe for consumption. There are several methods of desalination, each with its own advantages and limitations. Let's explore some of the most common techniques:

8.3.2.1 Distillation

Distillation is one of the oldest and most reliable methods of desalination. It involves heating seawater to create steam, which is then condensed and collected as freshwater. While distillation can effectively remove salt and other contaminants, it requires a heat source and can be energy-intensive.

8.3.2.2 Reverse Osmosis

Reverse osmosis is a popular desalination method that uses a semi-permeable membrane to separate salt and other impurities from seawater. The high-pressure system forces water through the membrane, leaving behind the salt and other contaminants. Reverse osmosis systems are widely used in coastal areas and on boats, as they are compact and efficient.

8.3.2.3 Solar Desalination

Solar desalination harnesses the power of the sun to evaporate seawater and collect the condensed freshwater. This method is particularly suitable for coastal and marine environments with abundant sunlight. Solar stills and solar desalination units are simple and cost-effective solutions for obtaining freshwater in these areas.

8.3.3 Collecting Rainwater in Coastal Areas

While desalination is an effective method for obtaining freshwater in coastal and marine environments, it is also important to explore alternative sources. One such source is rainwater. Although rainfall may be less frequent in coastal areas, it can still provide a valuable source of freshwater.

8.3.3.1 Maximizing Rainwater Collection

In coastal areas, it is crucial to maximize rainwater collection during the limited rainfall events. This can be achieved by using large catchment areas such as roofs, awnings, and other structures to collect as much rainwater as possible. Additionally, implementing gutter systems and downspouts can help direct the water into storage containers.

8.3.3.2 Filtering and Treating Rainwater

Rainwater, although generally considered safe for consumption, can still contain impurities and contaminants. It is important to filter and treat rainwater before using it for drinking or cooking. This can be done through various methods such as using sediment filters, activated carbon filters, and UV sterilization systems.

8.3.4 Fishing and Marine Life as a Water Source

Coastal and marine environments offer an abundance of marine life, which can serve as a source of water in times of need. Fishing and utilizing marine life for their water content can be a valuable survival skill. However, it is important to be knowledgeable about the local marine life and fishing regulations to ensure sustainability and avoid harmful practices.

8.4 Water in Urban Environments

In a crisis situation, urban environments present their own unique challenges when it comes to accessing clean and safe water. Unlike more rural areas, where natural water sources may be more readily available, urban areas often rely heavily on complex water infrastructure systems. When these systems fail or become compromised, it can quickly lead to a water scarcity crisis. As a prepper, it is crucial to understand the specific issues that arise in urban environments and be prepared to overcome them.

8.4.1 The Vulnerability of Urban Water Infrastructure

Urban areas are heavily dependent on centralized water treatment plants, pipelines, and distribution networks to provide clean and safe water to their residents. However, these systems are not infallible and can be vulnerable to various disruptions, such as natural disasters, power outages, or infrastructure failures. When these events occur, the supply of water to urban areas can be severely compromised, leaving residents without access to clean water.

8.4.2 Water Contamination Risks in Urban Environments

In addition to the vulnerability of water infrastructure, urban environments also pose unique risks of water contamination. The close proximity of industrial facilities, chemical storage sites, and densely populated areas increases the likelihood of pollutants entering the water supply during a crisis. Preppers in urban areas must be aware of these risks and take appropriate measures to ensure the water they consume is safe.

8.4.3 Alternative Water Sources in Urban Environments

When faced with a water scarcity crisis in an urban environment, it is essential to explore alternative water sources. While natural sources like rivers and lakes may be limited or contaminated, there are still options available. One such option is rainwater harvesting, which can be done even in urban settings with the use of rain barrels or rooftop collection systems. Additionally, exploring unconventional sources like condensation from air conditioning units or water from dehumidifiers can provide additional water resources.

8.4.4 Water Treatment in Urban Environments

In urban environments, where water infrastructure may be compromised, it is crucial to have effective water treatment methods in place. Boiling water, using chemical treatments, or employing filtration systems can help remove contaminants and make water safe for consumption. It is important to understand the specific challenges of treating water in an urban environment and choose the appropriate methods accordingly.

8.4.5 Water Storage Solutions for Urban Preppers

While storing large quantities of water in an urban environment may be challenging due to space constraints, it is still essential to have a sufficient supply of water for emergencies. Urban preppers can utilize various storage solutions, such as water storage containers, collapsible water bladders, or even repurposing household items like bathtubs or sinks. It is crucial to regularly rotate stored water to ensure its freshness and safety.

8.4.6 Community Collaboration for Water Security

In an urban environment, collaboration with neighbors and community organizations can be vital for ensuring water security during a crisis. Establishing neighborhood water committees, sharing resources, and pooling knowledge and skills can help overcome infrastructure issues and ensure a more reliable water supply for everyone. Building strong community networks and fostering a sense of collective responsibility can make a significant difference in urban water resilience.

8.4.7 Planning for Water Emergencies in Urban Environments

To effectively handle water-related emergencies in urban environments, it is crucial to have a well-thought-out emergency plan in place. This plan should include strategies for accessing alternative water sources, treating water, and storing water. It should also consider the specific challenges of the urban environment, such as limited space and potential contamination risks. Regularly reviewing and updating the plan will ensure its effectiveness when a crisis strikes.

In conclusion, water scarcity in urban environments can pose significant challenges during a crisis. Understanding the vulnerabilities of urban water infrastructure, exploring alternative water sources, implementing effective water treatment methods, and collaborating with the community are all essential for overcoming these challenges. By being prepared and having a well-defined emergency plan, urban preppers can ensure a reliable and safe water supply even in the most challenging circumstances.

Water for Food Production

9.1 Irrigation Techniques

In a crisis situation, having a reliable source of food is crucial for survival. And when it comes to growing your own food, water is the key ingredient that can make or break your efforts. In this chapter, we will explore various irrigation techniques that will help you water your survival garden efficiently and effectively, ensuring a bountiful harvest even in the most challenging circumstances.

9.1.1 Drip Irrigation: The Water-Saving Champion

When water is scarce, every drop counts. Drip irrigation is a highly efficient method that delivers water directly to the roots of your plants, minimizing water waste and maximizing plant growth. This technique involves using a network of tubes or pipes with small holes or emitters that release water slowly and steadily. By providing water directly to the root zone, drip irrigation reduces evaporation and ensures that plants receive the moisture they need without excess water being lost to the surrounding soil.

To set up a drip irrigation system, you will need a water source, such as a rain barrel or a storage tank, and a network of tubes or pipes with emitters. These emitters can be placed near the base of each plant or along the rows of your garden. By adjusting the flow rate and spacing of the emitters, you can customize the system to meet the specific water needs of different plants.

Imagine a scenario where water is scarce, and you have limited resources to grow your own food. With a well-designed drip irrigation system, you can make the most out of every drop of water, ensuring that your plants thrive even in challenging conditions. Not only does drip irrigation save water, but it also saves you time and effort by reducing the need for manual watering.

9.1.2 Sprinkler Irrigation: Covering a Larger Area

If you have a larger garden or need to cover a larger area, sprinkler irrigation can be a practical solution. This method involves using sprinklers to distribute water over the plants in a manner similar to rainfall. Sprinkler systems can be set up above ground or buried underground, depending on your specific needs and preferences.

Sprinkler irrigation is a versatile technique that allows you to adjust the water distribution pattern and intensity based on the requirements of different plants. By using different types of sprinkler heads, you can achieve a variety of spray patterns, including full-circle, half-circle, or even customized patterns to suit the shape and size of your garden.

One of the advantages of sprinkler irrigation is its ability to cover a larger area with minimal effort. With the right design and placement of sprinkler heads, you can ensure that every corner of your garden receives adequate water. However, it's important to note that sprinkler irrigation may not be as water-efficient as drip irrigation, as some water may be lost to evaporation or wind drift. Therefore, it's crucial to monitor and adjust your sprinkler system regularly to minimize water waste.

9.1.3 Flood Irrigation: Simplicity in Action

In situations where water resources are limited, and you need a simple and low-cost irrigation method, flood irrigation can come to the rescue. This technique involves flooding the garden beds or fields with water, allowing it to slowly seep into the soil and reach the plant roots.

Flood irrigation is a traditional method that has been used for centuries, and it requires minimal equipment and infrastructure. It can be particularly useful in areas with clay or loamy soils that have good water-holding capacity. By flooding the soil, you can ensure that water reaches the deeper layers, promoting root growth and providing a steady supply of moisture to your plants.

To implement flood irrigation, you will need a reliable water source and a system to control the flow of water. This can be as simple as using small ditches or furrows to direct the water to the desired areas. By monitoring the water level and adjusting the flow rate, you can ensure that your plants receive the right amount of water without causing waterlogging or runoff.

While flood irrigation may not be the most water-efficient method, it can be a viable option in certain situations, especially when water resources are limited, and you need a low-cost solution to keep your survival garden thriving.

9.1.4 Hydroponics: Growing Without Soil

In extreme water-scarce situations, traditional irrigation methods may not be feasible. That's where hydroponics comes into play. Hydroponics is a soilless gardening technique that allows you to grow plants in a nutrient-rich water solution, using only a fraction of the water required in traditional soil-based gardening.

In a hydroponic system, plants are grown in containers or trays filled with an inert growing medium, such as perlite or coconut coir. The roots of the plants are submerged or periodically sprayed with a nutrient solution that provides all the essential elements for growth. By directly delivering water and nutrients to the roots, hydroponics eliminates the need for excessive watering and reduces water waste.

Hydroponics offers several advantages in water-scarce environments. Firstly, it allows you to grow plants in a controlled environment, optimizing water and nutrient uptake. Secondly, it eliminates the risk of soil-borne diseases and pests, ensuring healthier plants and higher yields. Lastly, hydroponics can be implemented in small spaces, making it suitable for urban environments or areas with limited land availability.

By exploring hydroponics as an irrigation technique, you can unlock the potential to grow your own food even in the most challenging water conditions. With careful planning and the right equipment, you can create a self-sustaining hydroponic system that provides fresh produce year-round.

9.2 Aquaponics

Aquaponics is a revolutionary method of food production that combines fish farming (aquaculture) with hydroponics, creating a self-sustaining ecosystem that produces both fish and vegetables. In this chapter, we will explore the fascinating world of aquaponics and how it can be a valuable tool for preppers in ensuring a sustainable food supply in times of crisis.

9.2.1 The Basics of Aquaponics

Aquaponics is a symbiotic system where fish and plants work together to create a harmonious cycle of nutrient exchange. The fish produce waste, which is broken down by beneficial bacteria into nitrates. These nitrates serve as fertilizer for the plants, providing them with the essential nutrients they need to grow. In turn, the plants filter the water, removing harmful substances and providing a clean and oxygenated environment for the fish.

9.2.2 Setting Up an Aquaponics System

Setting up an aquaponics system may seem daunting at first, but with the right guidance, it can be a rewarding and relatively simple process. We will guide you through the step-by-step process of building your own aquaponics system, from choosing the right fish and plants to designing the layout and ensuring proper water circulation.

9.2.3 Choosing the Right Fish and Plants

The success of your aquaponics system depends on selecting the right combination of fish and plants that can thrive in the same environment. We will provide you with a comprehensive list of fish species that are well-suited for aquaponics, taking into consideration factors such as temperature tolerance, growth rate, and compatibility with other fish. Additionally, we will explore a variety of plant options, from leafy greens to herbs and even fruiting plants, and discuss their specific requirements for optimal growth.

9.2.4 Maintaining Your Aquaponics System

Like any other living system, aquaponics requires regular maintenance to ensure its smooth operation. We will guide you through the essential tasks of monitoring water quality, managing fish health, and maintaining the balance between fish and plant growth. We will also provide you with practical tips and tricks to troubleshoot common issues that may arise, such as nutrient deficiencies, pest control, and system malfunctions.

9.2.5 Maximizing Food Production with Aquaponics

Aquaponics offers a unique advantage in maximizing food production in limited space. We will explore various techniques and strategies to optimize your aquaponics system for maximum yield, including vertical gardening, companion planting, and staggered planting. By implementing these techniques, you can significantly increase the amount of fresh fish and vegetables you can produce, ensuring a sustainable food supply for you and your family.

9.2.6 Aquaponics in a Crisis Situation

In times of crisis, when access to traditional food sources may be limited, aquaponics can be a game-changer. We will discuss the advantages of aquaponics in emergency situations, such as its ability to operate off-grid, its low water consumption compared to traditional farming methods, and its potential for year-round food production. We will also provide practical advice on how to adapt your aquaponics system to function in a crisis, including alternative energy sources, water storage solutions, and emergency fish and plant selection.

9.2.7 Scaling Up and Community Aquaponics

Aquaponics is not limited to small-scale systems. We will explore the possibilities of scaling up your aquaponics operation to meet the needs of a larger community. By collaborating with like-minded individuals, you can create a network of aquaponics systems that can provide food security for a larger group of people. We will discuss the benefits and challenges of community aquaponics and provide guidance on how to establish and manage such a system.

9.3 Watering Livestock

Water is not only essential for human survival, but it is also crucial for the well-being and survival of livestock. In a crisis situation, when access to clean water becomes limited or compromised, it is vital for preppers to have the knowledge and skills to provide water for their animals. In this chapter, we will explore various methods and techniques for watering livestock, ensuring their health and productivity even in challenging circumstances.

9.3.1 Understanding the Water Needs of Livestock

Before we delve into the different methods of watering livestock, it is important to understand their water requirements. Just like humans, animals need water to survive, maintain their body temperature, and carry out essential bodily functions. The amount of water needed varies depending on the species, size, age, and environmental conditions. For example, a lactating cow may require up to 30 gallons of water per day, while a sheep may need around 2-4 gallons.

9.3.2 Natural Water Sources for Livestock

In a crisis situation, natural water sources such as rivers, lakes, and streams can be valuable resources for watering livestock. However, it is crucial to ensure that the water is safe for consumption. Contaminated water can lead to various health issues and even death in animals. Therefore, it is essential to have the knowledge and skills to identify and treat waterborne hazards, as discussed in Chapter 1.3.

9.3.3 On-Farm Water Storage Solutions

Having on-farm water storage solutions is a wise investment for preppers who own livestock. These storage solutions can help ensure a reliable and consistent water supply for animals, even during times of water scarcity. Rainwater harvesting, as discussed in Chapter 3, can be an effective method to collect and store water for livestock. Additionally, traditional water storage containers, such as tanks and troughs, can be used to store water for immediate use.

9.3.4 Watering Livestock in Arid Environments

In arid environments where water scarcity is a constant challenge, preppers must employ innovative techniques to ensure their livestock's water needs are met. One such technique is the use of water catchment systems, which collect and store rainwater from roofs or other surfaces. These systems can be connected to watering troughs or tanks, providing a reliable source of water for animals. Additionally, preppers can explore the use of underground water sources, such as wells or cisterns, to supplement their livestock's water requirements.

9.3.5 Portable Watering Solutions

In some situations, preppers may need to move their livestock to different locations to find suitable grazing areas or to escape from danger. In such cases, portable watering solutions can be invaluable.

Portable water troughs or tanks can be easily transported and set up in temporary grazing areas, ensuring that animals have access to water wherever they go. It is important to regularly monitor and replenish these portable water sources to meet the animals' needs.

9.3.6 Watering Livestock in Cold Climates

Cold climates present unique challenges when it comes to watering livestock. Freezing temperatures can cause water sources to freeze, making it difficult for animals to access water. Preppers must take measures to prevent water from freezing, such as using insulated troughs or installing heating elements. Additionally, breaking the ice regularly and providing warm water can help ensure that livestock stay hydrated during cold weather.

9.3.7 Water Quality and Animal Health

The quality of water provided to livestock directly impacts their health and productivity. Contaminated water can lead to various health issues, including diarrhea, dehydration, and reduced feed intake. Preppers should regularly test the water quality and take necessary measures to ensure it is safe for consumption. Water treatment methods, as discussed in Chapter 2, can be employed to purify water and eliminate harmful contaminants.

9.3.8 Emergency Watering Techniques

In emergency situations where access to water is severely limited, preppers may need to resort to alternative watering techniques. These techniques can include using water from unconventional sources, such as dew or plants, or utilizing water conservation methods to stretch the available water supply. It is important to prioritize the water needs of livestock and ensure they have access to clean water, even in challenging circumstances.

9.4 Preserving Food with Water

Preserving food is a crucial skill for any prepper. In a crisis situation, access to fresh food may become limited or even non-existent. That's why it's essential to learn how to preserve food for long-term storage. While there are various methods available, one often overlooked resource for food preservation is water. In this section, we will explore how water can be used to preserve food through canning and fermentation techniques.

9.4.1 Canning: A Time-Tested Preservation Method

Canning is a popular method of food preservation that has been used for centuries. It involves sealing food in jars or cans and then heating them to destroy any bacteria or microorganisms that could cause spoilage. The process creates a vacuum seal, preventing the entry of air and the growth of bacteria.

The Canning Process

To begin the canning process, you will need to gather the necessary equipment, including canning jars, lids, and a pressure canner or a boiling water bath canner. The food you wish to preserve should be prepared by washing, peeling, and cutting it into appropriate sizes. Next, you will pack the food into the jars, leaving the recommended headspace. The headspace allows for expansion during the canning process.

After packing the jars, you will need to add a liquid to the jar to ensure proper heat distribution during processing. Water is commonly used as the liquid, but you can also use fruit juices or syrups for added flavor. Once the jars are filled, you will need to remove any air bubbles by running a non-metallic utensil along the inside of the jar.

The next step is to place the lids on the jars and secure them with bands. For a boiling water bath canner, the jars are submerged in boiling water and processed for a specific amount of time. For a pressure canner, the jars are processed at a higher temperature and pressure. The processing time will vary depending on the type of food being canned.

Safety Considerations

It is crucial to follow proper canning procedures to ensure the safety of your preserved food. Botulism, a potentially deadly form of food poisoning, can occur if food is not canned correctly. Always use tested recipes and follow the recommended processing times and pressures. Inspect the jars for any signs of spoilage before consuming the preserved food.

9.4.2 Fermentation: Harnessing the Power of Microorganisms

Fermentation is another method of food preservation that relies on the action of beneficial microorganisms. These microorganisms convert sugars and carbohydrates in food into acids, alcohol, and gases, creating an environment that inhibits the growth of harmful bacteria.

The Fermentation Process

To begin the fermentation process, you will need to select the food you wish to ferment. Common examples include vegetables, fruits, and dairy products. The food is typically prepared by washing, peeling, and cutting it into the desired shape. It is important to use fresh and high-quality ingredients for successful fermentation.

Next, the food is placed in a container, such as a fermentation crock or a glass jar. A brine solution is then added to cover the food completely. The brine solution is made by dissolving salt in water. The salt acts as a natural preservative and creates an environment that favors the growth of beneficial bacteria while inhibiting the growth of harmful bacteria.

The container is then covered with a lid or a cloth to allow for the release of gases produced during fermentation. The food is left to ferment at room temperature for a specific period, which can range from a few days to several weeks, depending on the desired flavor and texture.

Safety Considerations

While fermentation is generally a safe method of food preservation, it is essential to follow proper techniques to prevent the growth of harmful bacteria. Use clean utensils and containers to avoid contamination. Monitor the fermentation process closely and discard any food that shows signs of spoilage, such as mold or off-putting odors.

Preparing for Long-Term Water Scarcity

10.1 Drought Preparedness

Droughts can be devastating, leaving communities and individuals without access to the water they need to survive. As a prepper, it is crucial to be prepared for extended water shortages and have a plan in place to cope with the challenges that droughts bring. In this chapter, we will explore the importance of drought preparedness and provide you with practical strategies to ensure you have enough water to sustain yourself and your loved ones during these difficult times.

Understanding the Impact of Droughts

Droughts occur when there is a prolonged period of abnormally low rainfall, resulting in a shortage of water. These dry spells can have severe consequences, affecting agriculture, wildlife, and human populations. Understanding the impact of droughts is essential for developing effective strategies to mitigate their effects.

Developing a Drought Preparedness Plan

To ensure your survival during a drought, it is crucial to develop a comprehensive drought preparedness plan. This plan should include strategies for conserving water, finding alternative water sources, and storing water for long-term use. Let's explore these strategies in more detail.

Water Conservation Techniques
Conserving water is the first line of defense during a drought. By reducing your water usage, you can stretch your available water supply and minimize the impact of the drought on your daily life. Some practical water conservation techniques include:

- Fixing leaky faucets and pipes to prevent water wastage.

- Installing low-flow showerheads and toilets to reduce water consumption.

- Collecting and reusing greywater for non-potable purposes such as flushing toilets or watering plants.

- Limiting outdoor water usage by watering plants during cooler hours and using drip irrigation systems.

Finding Alternative Water Sources
When traditional water sources run dry during a drought, it becomes essential to find alternative sources of water. This may include exploring nearby rivers, lakes, or streams that are still flowing, or tapping into underground water sources such as

wells or springs. However, it is crucial to ensure the safety of these alternative water sources by properly treating and purifying the water before consumption.

Storing Water for Long-Term Use
Having a reliable and sufficient water storage system is vital during a drought. It is recommended to store at least one gallon of water per person per day for a minimum of two weeks. This ensures that you have enough water for drinking, cooking, and personal hygiene. Consider using food-grade water storage containers and regularly rotating your water supply to keep it fresh.

10.2 Water Conservation in the Long Run

Water conservation is not just a short-term solution to cope with water scarcity during a crisis; it is a sustainable practice that should be embraced by preppers for the long run. As a prepper, you understand the vital importance of water and the need to be prepared for water-related emergencies. In this chapter, we will explore the various ways you can conserve water to ensure a sustainable water supply for yourself and your community.

10.2.1 The Value of Water Conservation

Water conservation is not just about saving water; it is about preserving a precious resource that is essential for life. By conserving water, you not only ensure your own survival but also contribute to the well-being of the environment and future generations. In times of crisis, when water sources may be limited or contaminated, practicing water conservation becomes even more critical.

10.2.2 Practical Water Conservation Techniques

In this section, we will delve into practical water conservation techniques that you can implement in your daily life. From simple changes in your habits to more advanced strategies, these techniques will help you reduce water waste and maximize the efficiency of your water usage.

10.2.2.1 Fixing Leaks and Drips

One of the most common sources of water waste is leaks and drips in our plumbing systems. A small leak may seem insignificant, but over time, it can waste a significant amount of water. We will discuss how to detect and fix leaks in your pipes, faucets, and toilets, ensuring that every drop of water is put to good use.

10.2.2.2 Efficient Water Appliances

Upgrading to water-efficient appliances can make a significant difference in your water consumption. We will explore the latest technologies in water-saving appliances such as low-flow toilets, showerheads, and dishwashers. Additionally, we will provide tips on how to use these appliances effectively to maximize water conservation.

10.2.2.3 Outdoor Water Conservation

Conserving water extends beyond the walls of your home. In this section, we will discuss strategies for conserving water in your outdoor spaces, such as your garden and lawn. From using drought-resistant plants to implementing smart irrigation systems, we will provide you with practical tips to minimize water usage while maintaining a beautiful outdoor environment.

10.2.2.4 Rainwater Harvesting and Greywater Recycling

Rainwater harvesting and greywater recycling are effective ways to conserve water and reduce reliance on traditional water sources. We will delve into the details of these techniques, explaining how to set up rainwater collection systems and safely reuse greywater for non-potable purposes.

10.2.3 Community Water Conservation

Water conservation is not just an individual effort; it is a collective responsibility. In this section, we will explore the importance of community involvement in water conservation. We will discuss the benefits of organizing community water conservation initiatives, such as water-saving campaigns, educational programs, and infrastructure improvements. By working together, communities can create a more resilient and sustainable water future.

10.2.4 Planning for Long-Term Water Security

To ensure long-term water security, it is crucial to have a comprehensive water emergency plan in place. We will guide you through the process of creating a personalized plan that addresses your specific needs and circumstances. From assessing your water requirements to identifying alternative water sources and storage solutions, we will provide you with the tools and knowledge to prepare for extended water shortages.

10.3 Community Water Solutions

Water scarcity is a serious concern, especially in times of crisis or emergency. As a prepper, it is crucial to not only prepare yourself but also your community for potential water-related emergencies. In this chapter, we will explore various community water solutions that can help ensure the resilience and survival of your community in times of water scarcity.

10.3.1 Establishing a Community Water Committee

One of the first steps in preparing for long-term water scarcity is to establish a community water committee. This committee should consist of individuals who are knowledgeable about water management, treatment, and conservation. By working together, the committee can develop strategies and plans to address water-related challenges in your community.

10.3.2 Rainwater Harvesting Projects

Rainwater harvesting is an effective way to collect and store water for future use. By implementing rainwater harvesting projects in your community, you can ensure a sustainable and reliable water source. This could involve setting up rainwater collection systems on rooftops, constructing storage tanks, and implementing purification methods to make the harvested rainwater safe for consumption.

10.3.3 Community Water Treatment Centers

In times of crisis, access to clean and safe drinking water becomes even more critical. Establishing community water treatment centers can help ensure that everyone in your community has access to treated water. These centers can be equipped with filtration systems, chemical treatment methods, and trained personnel to ensure the water is safe for consumption.

10.3.4 Water Conservation Campaigns

Water conservation is a key aspect of preparing for long-term water scarcity. By organizing water conservation campaigns in your community, you can raise awareness about the importance of water conservation and encourage everyone to adopt water-saving practices. These campaigns can include educational workshops, distribution of water-saving devices, and community-wide initiatives to reduce water waste.

10.3.5 Community Water Sharing Programs

In times of crisis, some members of your community may face more significant water shortages than others. Implementing community water sharing programs can help ensure that everyone has access to an adequate water supply. These programs can involve the distribution of water among community members, prioritizing those in need, and establishing guidelines for fair and equitable water sharing.

10.3.6 Community Water Infrastructure Projects

Investing in community water infrastructure projects can significantly improve water resilience in your community. This could include the construction of wells, cisterns, or water storage facilities that can provide a reliable water source during times of scarcity. By working together as a community, you can pool resources and expertise to implement these infrastructure projects effectively.

10.3.7 Education and Training Programs

Educating and training community members on water management, treatment, and conservation is crucial for long-term water resilience. By organizing workshops, seminars, and training programs, you can empower individuals in your community to take an active role in water-related emergencies. This knowledge can help them make informed decisions, implement effective water-saving practices, and contribute to the overall water security of the community.

10.3.8 Collaboration with Local Authorities and Organizations

Collaborating with local authorities and organizations is essential for the success of community water solutions. By partnering with government agencies, non-profit organizations, and other community groups, you can access additional resources, expertise, and funding to implement water projects effectively. Together, you can work towards creating a more resilient and water-secure community.

10.3.9 Monitoring and Evaluation

Regular monitoring and evaluation of community water solutions are crucial to ensure their effectiveness and sustainability. By establishing monitoring systems, you can track the progress of your community water projects, identify areas for improvement, and make necessary adjustments. This ongoing evaluation will help you refine your strategies and ensure the long-term success of your community water solutions.

10.3.10 Celebrating Success and Sharing Knowledge

As your community implements various water solutions, it is important to celebrate successes and share knowledge with other communities facing similar challenges. By sharing your experiences, lessons learned, and best practices, you can inspire and support other communities in their journey towards water resilience. Together, we can build a network of communities prepared to handle water-related emergencies.

In conclusion, community water solutions are vital for ensuring the resilience and survival of your community in times of water scarcity. By establishing a community water committee, implementing rainwater harvesting projects, setting up water treatment centers, organizing water conservation campaigns, and collaborating with local authorities and organizations, you can create a more water-secure community. Remember, working together as a community is the key to overcoming water-related challenges and ensuring the well-being of everyone in times of crisis.

10.4 Planning for Water Security

Water is the essence of life, and in times of crisis, it becomes even more precious. As a prepper, you understand the importance of being prepared for any emergency, and that includes having a plan in place for water security. In this chapter, we will delve into the crucial aspects of creating a personal water emergency plan to ensure that you and your loved ones have access to safe and sufficient water during times of scarcity.

10.4.1 Assessing Your Water Needs

Before you can create a water emergency plan, it's essential to assess your water needs. Start by considering the number of people in your household and their daily water consumption. Take into account factors such as drinking, cooking, hygiene, and sanitation. Remember that during emergencies, water needs may increase due to stress, physical exertion, and the lack of alternative water sources.

To help you determine your water needs, here's a simple calculation:

- Multiply the number of people in your household by the recommended daily water consumption, which is generally around one gallon (3.8 liters) per person.

- Add extra water for cooking, hygiene, and sanitation purposes. A good rule of thumb is to double the daily water consumption to account for these additional needs.

Once you have calculated your daily water needs, multiply that number by the number of days you want to be prepared for. It's recommended to have at least a two-week supply of water for emergencies, but you may choose to store more depending on your circumstances.

10.4.2 Water Storage and Rotation

Now that you know how much water you need, it's time to consider how to store and rotate your water supply. Proper water storage is crucial to ensure that your water remains safe to drink over an extended period. Here are some key points to keep in mind:

- Choose food-grade containers that are specifically designed for water storage. These containers should be made of materials that won't leach harmful chemicals into the water.

- Clean and sanitize your containers before filling them with water. Use a mixture of one teaspoon of unscented household bleach per gallon of water to disinfect the containers.

- Store your water in a cool, dark place away from direct sunlight. Sunlight can promote the growth of algae and bacteria in your water supply.

- Regularly inspect your water containers for any signs of damage or leakage. Replace any damaged containers immediately.

- Rotate your water supply every six months to ensure freshness. Use the oldest water for non-potable purposes and refill your storage containers with fresh water.

10.4.3 Water Sources and Treatment

While having a stored water supply is essential, it's also crucial to have a plan for finding and treating water from alternative sources. During a long-term water scarcity situation, your stored water may eventually run out, and you'll need to rely on other sources. Here are some key considerations:

- Identify potential water sources in your area, such as rivers, lakes, and streams. Research their accessibility and proximity to your location.

- Learn about different water treatment techniques, such as boiling, chemical treatment, and filtration. Understand the pros and cons of each method and have the necessary equipment and supplies on hand.

- Consider investing in a portable water filter or purifier that can effectively remove contaminants from natural water sources.

- Familiarize yourself with local plants and their water content. In some cases, you may be able to extract water from plants through condensation or other methods.

- Develop skills for rainwater harvesting and collection. This can be an excellent source of water during rainy seasons.

10.4.4 Emergency Water Conservation

In times of water scarcity, it's crucial to conserve water to make your supply last longer. Here are some practical tips for emergency water conservation:

- Limit your water usage by taking shorter showers and turning off the tap while brushing your teeth or washing dishes.

- Collect and reuse greywater from activities such as dishwashing or laundry for non-potable purposes like flushing toilets or watering plants.

- Fix any leaks in your plumbing system promptly. Even small leaks can waste a significant amount of water over time.

- Prioritize essential water needs and find alternative ways to meet non-essential needs. For example, consider using dry shampoo or wet wipes for personal hygiene when water is scarce.

By incorporating these water conservation practices into your daily life, you'll not only extend the lifespan of your stored water but also develop habits that promote long-term sustainability.

10.4.5 Emergency Water Treatment and Purification

In addition to having a plan for finding alternative water sources, it's crucial to know how to treat and purify water to make it safe for consumption. Here are some key points to consider:

- Understand the different water treatment methods, such as boiling, chemical treatment, filtration, and solar disinfection. Each method has its advantages and limitations, so it's essential to be familiar with all of them.

- Have the necessary equipment and supplies for water treatment readily available in your emergency kit. This includes water filters, water purification tablets, and portable water purifiers.

- Learn how to properly use and maintain your water treatment equipment to ensure its effectiveness.

- Stay informed about potential waterborne hazards in your area, such as bacteria, viruses, and chemical contaminants. This knowledge will help you choose the appropriate treatment method for different situations.

Remember, water treatment is a critical step in ensuring the safety of your water supply. By being prepared and knowledgeable about water treatment techniques, you can protect yourself and your loved ones from waterborne illnesses during emergencies.

10.4.6 Creating a Water Emergency Plan

Now that you have a good understanding of the various aspects of water security, it's time to put it all together and create a comprehensive water emergency plan. Here's a step-by-step guide to help you get started:

- Assess your water needs based on the number of people in your household and their daily water consumption.

- Determine how much water you need to store for your desired duration of preparedness.

- Choose appropriate water storage containers and ensure they are clean and in good condition.

- Store your water in a cool, dark place away from direct sunlight.

- Regularly inspect and rotate your water supply to ensure freshness.

- Identify potential alternative water sources in your area and learn how to treat and purify water from these sources.

- Develop water conservation habits to make your water supply last longer.

- Familiarize yourself with different water treatment methods and have the necessary equipment and supplies on hand.

- Stay informed about potential waterborne hazards in your area and adjust your treatment methods accordingly.

- Review and update your water emergency plan regularly to account for any changes in your circumstances or local conditions.

By following these steps and incorporating them into your overall emergency preparedness strategy, you can ensure that you and your loved ones have a reliable and secure water supply during times of crisis.

Remember, water is a precious resource, and being prepared for water-related emergencies is not only a matter of survival but also a responsible and caring approach to protecting yourself and your community.

Conclusion

As we draw this guide to a close, it's clear that this book stands as an indispensable resource for preppers and anyone committed to being well-prepared for water-related emergencies. This book has meticulously covered the indispensable role water plays in survival scenarios, highlighting the critical need to comprehend various water sources, potential contaminants, and the essential techniques for securing water in adverse conditions.

Through the chapters, we've embarked on an in-depth exploration of numerous water purification methods, including boiling, chemical treatments, filtration, and the innovative practice of solar disinfection. Additionally, the book has shed light on the principles of rainwater harvesting, the nuances of water conservation, identifying emergency water reserves, and the pivotal practices for effective water storage and rotation.

The discussion extended beyond mere survival to encompass water's significance in maintaining hygiene, sanitation, and its crucial role in food production, addressing the hurdles posed by harsh environments and the specter of prolonged water scarcity. This comprehensive discourse has armed us with the knowledge and practical skills vital for confronting water-related predicaments with confidence and resilience.

In an era marked by growing concerns over water scarcity and emergency situations, this book serves as a powerful tool, encouraging readers to proactively establish a water emergency strategy. It guides the assessment of water needs, elucidates various purification techniques, and advocates for the adoption of water conservation measures. This proactive approach ensures the readiness to provide safe, adequate water supplies for oneself and loved ones.

Ultimately, water is the lifeline of existence. Preparing for water-related emergencies is an act of foresight, embodying a commitment to safeguarding our well-being and that of our communities. Equipped with the insights and strategies detailed in "The Prepper's Water Survival Bible," we are empowered to tackle water challenges head-on. This knowledge ensures that we are not just survivors but resilient protectors of our most precious resource. Let us move forward with preparedness and assurance, secure in the knowledge that our water needs will be met, come what may. Stay vigilant, stay safe, and ensure your water security is unassailable.

www.ingramcontent.com/pod-product-compliance
Lightning Source LLC
Chambersburg PA
CBHW071052240526
45471CB00015B/1643